THE
GOOD
SOCIETY

BOOKS BY
JOHN KENNETH GALBRAITH

American Capitalism:
The Concept of Countervailing Power

A Theory of Price Control

The Great Crash, 1929

The Affluent Society

The Scotch

The New Industrial State

The Triumph

Indian Painting
(with Mohinder Singh Randhawa)

Ambassador's Journal

Economics and the Public Purpose

Money: Whence It Came, Where It Went

Annals of an Abiding Liberal

A Life in Our Times

The Anatomy of Power

A View from the Stands

Economics in Perspective:
A Critical History

A Tenured Professor

The Culture of Contentment

A Journey Through Economic Time:
A Firsthand View

The Good Society:
The Humane Agenda

John Kenneth Galbraith

THE
GOOD
SOCIETY

The Humane Agenda

HOUGHTON MIFFLIN COMPANY

Boston • New York

For information about permission to reproduce selections from
this book, write to Permissions, Houghton Mifflin Company,
215 Park Avenue South, New York, New York 10003.

For information about this and other Houghton Mifflin trade
and reference books and multimedia products, visit
The Bookstore at Houghton Mifflin on the World Wide Web
at http://www.hmco.com/trade/.

Library of Congress Cataloging-in-Publication Data
Galbraith, John Kenneth, date.
The good society : the humane dimension / John
Kenneth Galbraith.
p.   cm.
Includes index.
ISBN 0-395-71328-5
1. Welfare economics. 2. Income distribution. 3. Social
justice. 4. Consumption (Economics) 5. Individualism.
I. Title
HB846.G35   1996   96-983
330.12'6 — dc20   CIP

Printed in the United States of America
Book design by Robert Overholtzer

EB 10 9 8 7 6 5

ACKNOWLEDGMENTS

My first word of thanks goes to the Deutsche Evangelische Kirche (the biyearly meeting of the German Evangelical Church), which in the early summer of 1993 gathered the many thousands in Munich and asked me to speak on the good society. This started a current of thought and effort that I then pursued, as other obligations allowed, for the next two years, strengthened, as I later tell, by recent political developments and deviance in the United States and elsewhere.

The Good Society has been used as a title on various works before, and with no slight popular effect on a treatise by Walter Lippmann in 1937. There was no search for imitative distinction here. *The Good Society* merely expresses with the greatest clarity my intention in this exercise.

As ever, I thank my Harvard colleagues with whom I have discussed these matters and my son James Galbraith, professor at the University of Texas, who has given me access to his excellent computer bank. Andrea Williams, my friend and collaborator for thirty-seven years, has, as before, brought to bear her editorial skills, her good humor

Acknowledgments

and a certain patient persistence developed over the decades so that my English prose does not arouse the concern or the compassion of my critics. To Andrea, truly my thanks. Brooke Palmer, my very effective administrative assistant, has with tact and skill fended off or absorbed competing claims on my time, making it possible for me to write and, I trust, to think. I have a special word for my publisher, Houghton Mifflin Company, with whom I have also had a friendly association for almost half a century. Rarely have author and publisher combined so agreeably for so long.

Finally, and certainly not least, Catherine Atwater Galbraith has, as so often, been my beloved and wholly tolerant supporter in the writing of this book. It was fitting that my original inspiration should have occurred in Munich, for it was there as a graduate student that she had a significant part of her own scholarly career. Ever since a sparkling day in the autumn of 1937, she has watched over all my efforts with patience, encouragement and loving tolerance. To Kitty especially, my thanks and my love.

JOHN KENNETH GALBRAITH
Cambridge, Massachusetts
November 1995

CONTENTS

For Sissela and Derek Bok

1

The Good Society

AMONG THE GREAT NATIONS of the world none is more given to introspection than the United States. No day passes without reflective comment — by the press, on radio or television, in an article or book, in compelled and sometimes compelling oratory — on what is wrong in the society and what could be improved. This is also, if in lesser measure, a preoccupation in the other industrial lands — Britain, Canada, France, Germany, elsewhere in Europe and in Japan. No one can deplore this exercise; far better and far more informative such a search than the facile assumption that all is well. Before knowing what is right, one must know what is wrong.

There is, however, another, less traveled course of thought. That is to explore and define what, very specifically, would be right. Just what should the good society be? Toward what, stated as clearly as may be possible, should we aim? The tragic gap between the fortunate and the needful having been recognized, how, in a practical way, can it be closed? How can economic policy contribute to this end? What of the public services of the state; how can they be made more equitably and efficiently available? How can the environment, present and future, be protected? What of immigration, migration and migrants? What of the military power? What is the responsibility and course of action of the good society as regards its trading partners and neighbors in an increasingly internationalized world and as regards the poor of the planet? The responsibility for economic and social well-being is general, transnational. Human beings are human beings wherever they live. Concern for their suffering from hunger, other deprivation and disease does not end because those so afflicted are on the other side of an international frontier. This is the case even though no elementary truth is so consistently ignored or, on occasion, so fervently assailed.

To tell what would be right is the purpose of this book. It is clear at the outset that it will encounter a difficult problem, for a distinction must be made, a line drawn, between what might be perfect and what is achievable. This task and the result may not be politically popular and certainly not in a polity where, as I shall argue, the fortunate are now socially and politically dominant. To identify and urge the good and achievable society may well be a minority effort, but better that effort than none at all. Perhaps, at a minimum, the comfortable will be afflicted in a useful way. In any case, there is no chance for the better society unless the good and achievable society is clearly defined.

It is the achievable, not the perfect, that is here identified and described. To envision a perfect society has not in the past been an unattractive exercise; over the centuries it has been attempted by many scholars and not a few of the greatest philosophers. It is also, alas, a formula for dismissal. The predictable reaction is the statement that one's goals are "purely utopian." The real world has constraints imposed by human nature, by history and by deeply ingrained patterns of thought. There are also constitutional restraints and other long-established legislative procedures as well as the controls attendant on the political party system. And there is the fixed institutional structure of the economy — the corporations and the other business enterprises, large and small, and the limits they impose. In all the industrial countries, there is the firm commitment to the consumer economy — to consumer goods and services — as the primary source of human satisfaction and enjoyment and as the most visible measure of social achievement. There is also the even more urgent need for the income that comes from production. In the modern economy, a slightly bizarre fact, production is now more necessary for the employment it provides than for the goods and services it supplies.

Any useful identification of the good society must therefore take into consideration the institutional structure and the human characteristics that are fixed, immutable. They make the difference between the utopian and the achievable, between the agreeably irrelevant and the ultimately possible.

To define the achievable is the most difficult problem with which an essay such as this contends. It is also the most controversial. To call some urgently required action politically or socially impossible is the first (and sometimes the only) line of defense against unwanted change.

This book tells of the good society that is the achievable society. It accepts that some barriers to achievement are immobile, decisive, and thus must be accepted. But there are also goals that cannot be compromised. In the good society all of its citizens must have personal liberty, basic well-being, racial and ethnic equality, the opportunity for a rewarding life. Nothing, it must be recognized, so comprehensively denies the liberties of the individual as a total absence of money. Or so impairs it as too little. In the years of Communism it is not clear that one would wisely have exchanged the restraints on freedom of the resident of East Berlin for those imposed by poverty on the poorest citizens of the South Bronx in New York. Meanwhile, nothing so inspires socially useful effort as the prospect of pecuniary reward, both for what it procures and not rarely for the pleasure of pure possession it accords. This too the good society must acknowledge; these motivations are controlling.

As there are shaping forces, some deep in human nature, that must be accepted, so there are constraints that the good society cannot, must not, accept. Socially desirable change is regularly denied out of well-recognized self-interest. In the most important current case, the comfortably affluent resist public action for the poor because of the threat of increased taxes or the failure of a promise of tax reduction. This, the good society cannot accept. The seemingly decisive constraint here, in fact, is a political attitude that supports and sustains the very conditions that require correction. When it is said that some action may be good but is politically impractical, it must be understood that this is the common design for protecting a socially adverse interest.

It is the nature of privileged position that it develops its own political justification and often the economic and social doctrine that serves it best. No one likes to believe that his or her personal well-being is in conflict with the greater public need. To invent a plausible or, if necessary, a moderately implausible ideology in defense of self-interest is thus a natural course. A corps of willing and talented craftsmen is available for the task. And such ideology gains greatly in force as those who are favored increase in number. The pages that follow contend with but do not respect this broad tendency. Their purpose is to challenge it wherever, as is often the case, it stands against the larger and more urgent public need.

2

The Wider Screen

IN A BOOK published a few years ago,* I observed that in the rich countries of the world, and notably in the United States, there was a new political dialectic. Once there was employer versus employed; the capitalist, great and less great, versus the working masses, the latter in varying relationship with landlords, peasantry and in the United States the independent farmers. There was always effort to put the opposing interests in benign terms: the system as a whole served the interest of all; the overriding role of constitutional democracy protected liberties and ensured a reasonably peaceful resolution of the inherent differences; everything was for the best.

There was, nonetheless, conflict implicit in all reputable economic and political thought. It shaped the development of modern politics in the United States, Western Europe and Japan. On the one hand, there were the liberals, as they were called in the United States; the socialists and social democrats, as elsewhere they were named;

*The Culture of Contentment (Boston: Houghton Mifflin, 1992)

on the other, asserting or accepting the business interest, the conservatives. There were many variants in practical politics, concessions from one side to the other, often reluctantly extracted. Wider issues — peace and war, religious commitment, ethnic and racial equality — intruded. In the United States a large rural population helped to mellow the conflict. Always present, however, was the basic, the ultimate, dichotomy: capital versus labor. That, to repeat, was assumed in all political discourse and action.

Now it can be assumed no longer. The old dichotomy survives in the public psyche — the residue of its long and ardent history. But in the modern economy and polity the division is very different, and this is so in all the economically advanced countries. On one side, there are now the rich, the comfortably endowed and those so aspiring, and on the other the economically less fortunate and the poor, along with the considerable number who, out of social concern or sympathy, seek to speak for them or for a more compassionate world. This is the economic and political alignment today.

The rich and the well situated are now far more numerous and diverse than the erstwhile capitalist class, and they are also politically much more articulate. (The great capitalists were often slightly reticent as to their public role and interest.) The less favored are the poverty-stricken of the great cities, those who staff the service industries, the unemployable and the unemployed. And those who suffer from residual racial, gender or age discrimination or who are recent and sometimes illegal immigrants. All are largely without political voice except as they are supported and represented by the considerable number in the more fortunate brackets who feel and express concern.

Here in briefest form is the modern political dialectic. It is an unequal contest: the rich and the comfortable have influence and money. And they vote. The concerned and the poor have numbers, but many of the poor, alas, do not vote. There is democracy, but in no slight measure it is a democracy of the fortunate.

A defining issue between these two groupings, as is well recognized, is the role of government. For the poor, the government can be central to their well-being, and for some even to survival. For the rich and the comfortable, it is a burden save when, as in the case of military expenditure, Social Security and the rescue of failed financial institutions, it serves their particular interest. Then it ceases to be burdensome and becomes a social necessity, a social good, as certainly it is not when the government serves the poor.

In the congressional and state elections in the autumn of 1994, there was a massive swing to the political right in the United States. The principal issue was the just mentioned role of government and its cost, always with the exceptions already noted. The victory was not quite as significant in quantitative terms as has sometimes been suggested. Fewer than half of those eligible went to the polls; the prevailing candidates won with slightly less than one quarter of the eligible vote. While *The Good*

Society had been under way for some time prior to the election, the outcome of the latter sharply affirmed the purpose of the book, which is to state in as clear terms as possible what should be the goal not for the fortunate but for all.

This may now seem outside the limits of the achievable as they were earlier discussed. One may be sure that those who define politics in terms of the seemingly practical will so believe and certainly so say. The trend of the time is in the opposite direction. Let romance not disguise reality: in the United States one influential part of the media defines as truth the currently popular political attitude.

This is to ignore a far deeper truth — to fail to appreciate the more fundamental thrust of history, which is greater than current action and reaction and has a controlling influence of its own. It is the pride of liberals and the political conviction of conservatives that they shape the social agenda; in fact, it is shaped by the deeper trends of history. To these there must be accommodation, and liberals, social democrats and those called socialists in the advanced countries have traditionally made or guided this accommodation. In consequence, to them has been attributed the larger change; some, indeed many, have taken credit therefor, and conservatives have all but universally awarded them responsibility and blame. But, in reality, it is history that is in control. The briefest look at the basic circumstances readily establishes the point.

Until the early decades of the present century the United States was predominantly a rural country. As late as the

9

Great Depression, approaching half of all gainfully employed workers were in agriculture. Many more were in small-scale merchant, service and other village enterprises. In this economic and social context there was no urgent need for Social Security, one of the great transforming steps of the time, for here the next generation looked after the last. Or from the sale of the farm or small business came the wherewithal, life expectancy being what it was, to support the relatively brief retirement. It was the longer life span provided by modern medicine but, more important, the rise of urban industry and employment, not liberals or socialists, that created the pressure for Social Security.

It was also industrial and urban development that made unemployment a problem. In traditional agriculture it did not exist; there was always work to do on the farms and in the supporting rural services. (In the Depression farm employment or farm existence of a sort was the resort of some millions of urban workers in the United States.) Because of industrial development and urbanization, unemployment compensation became essential.

Modern medical insurance is also the offspring of history. Until relatively recent times medical knowledge was limited, as was the possibility of remedial effect. The local doctor had little to sell; death was early, inevitable and inexpensive. It was the enormous growth and improvement in medical and surgical procedures that made health insurance both necessary and desirable. This was the ultimate and motivating force. Death would no longer be the automatic prognosis for the poor or the only moderately affluent.

The simple living standard of earlier and by no means distant times posed few problems as to product safety or reliability. Basic foods, clothing, house room could all be appraised quite adequately by the purchaser; no deeper information was required. Until recently agriculture and elementary industry had little adverse environmental effect, and neither did their marketing outlets and suppliers. Now, with the expansion and complexity of the economy, consumers must be protected, along with the environment.

But there is more. The poor in the United States, while none could doubt their degradation and misery, were once largely invisible — poor blacks were hidden away on the farms and plantations of the rural South with primitive food, clothing and shelter, little in the way of education and no civil rights. Many poor whites were unseen on the hills and in the hollows of Appalachia. Poverty was not a problem when distant, out of sight. Only as economic, political and social change brought the needy to the cities did welfare become a public concern, the poor now living next to and in deep contrast with the relatively affluent.

The force of history extends to foreign policy. Before the United States became a world power, the Department of State was a small, comfortable enclave of well-bred gentlemen pursuing an effortless routine of no great consequence. It was only with the emergence of the United

States as a major player on the international scene and the breakup of the colonial world, leading on to the problems and the conflicts of the poor countries, to the question of economic assistance and the more than occasional necessity for intervention to restore peace and tranquillity, that foreign policy became a major preoccupation.

Here then is the error: in the common view of both liberals and conservatives in the United States, it is the liberals who have made government a large, intrusive force. Both groups wish to believe that it is political decision and action that are controlling. And from this comes the prime conservative notion that social and economic policy can be reversed, a view held not alone in the United States but in France, Canada and for long years in Britain, where there is or has been a similar belief among the Tories.

History, the truly relevant source of change, will not be reversed. The new Congress that came to office in the United States in early 1995 representing the conservative will expressed its intention to dismantle much of the welfare state, much of the modern regulatory apparatus of government, and to limit drastically the role of government in general. This was the broad promise broadly enunciated. Then came the specific legislation, the assault on particular functions and regulations. As this is written, these are proving far from popular; once more we see the not unusual conflict between broad theory and specific action. Some dramatic and well-publicized exceptions possibly apart, the welfare state and its basic programs will survive. The larger force of history is still at work.

The public and political actions to be proposed in these pages are in keeping with the controlling forces just cited, and consistent with them there is much to be urged, much to be done. The accommodation to historical trends can be improved, made more compassionate in order to provide a better life for the more vulnerable elements. This, to repeat, is the theme of what will follow. There are two questions to be answered. Within the larger historical framework, what is the nature of the good society? How can the future be made safer and better for all?

3

The Age of
Practical Judgment

NCIENTLY AND STILL, the economy has been
defined ideologically. There is liberalism, socialism or capitalism; the speaker is a liberal or a
socialist or is for free enterprise. He or she favors public
ownership or, as in recent times, privatization. These are
the controlling rules within which we live.

There is in the present day no greater or more ardently
argued error. In the modern economic and political system ideological identification represents an escape from
unwelcome thought — the substitution of broad and ba-
nal formula for specific decision in the particular case. A
look at the most elementary of present circumstance
proves the point.

An evident purpose of the good economy is to produce
goods and render services effectively and to dispense the
revenues therefrom in a socially acceptable and economically functional manner. There can be no question that
the modern market economy in the economically advanced countries does produce consumer goods and serv-

ices in a competent, even lavish fashion. Not only does it supply food, clothing, furniture, automobiles, entertainment and much else in diverse abundance, but it goes far to create the wants that it so satisfies. The sovereignty of the consumer is one of the most cherished ideas in orthodox economics; that this sovereignty has, in substantial degree, been surrendered to those who serve it is the most resisted. Yet nothing is more apparent than modern advertising and merchandising effort. Economists committed to the more rigorous levels of accepted thought do not watch television.

It therefore defies all sense that the supply of consumer goods and services, this lush operation, should somehow be taken over by the state. The revelation by television and other modern communications of the manifest abundance and variety of material possessions in the Western countries was one factor unsettling the socialist regimes of Eastern Europe and the former Soviet Union. The weakness and rigidity with which they had supplied their citizens with such goods and services in the required quantity, styles and changing fashions had more than a little to do with their downfall. To speak for socialism, public ownership, in the consumer-goods economy verges on the fanciful, and it is equally fanciful to urge the case on the producers of the plant and equipment — the capital goods — that manufacture this consumer abundance.

The traditional argument for socialism had a deeper claim on public attention. It revolved around the possession of power, and this remains important in some cellular recesses of social thought to this day. The private ownership of capital, of the means of production; the mass of workers thus employed and in great measure so controlled; the personal wealth resulting; the intimate association with the state; did once accord decisive power. Of this there can be no doubt. Marx and Engels, in *The Communist Manifesto*, said with no great exaggeration that "the executive of the modern state is but a committee for managing the common affairs of the whole bourgeoisie."

It is not doubted that power still resides with the ownership of capital. But in the enormous business enterprises of today ownership and control are, in the normal case, no longer united. The great capitalist entrepreneurs who both owned and commanded — Vanderbilt, Rockefeller, Morgan, Harriman in the United States and their counterparts in other countries — are gone forever. In their place is the massive, often immobile corporate bureaucracy and the financially interested but functionally ineffective stockholders. Monopoly power — exploitation of the consumer by prices unrestrained by competition and once in the United States the object of the antitrust laws — has surrendered to international competition and also to explosive technological change. Today's eminence and economic influence are tomorrow's obsolescence. Replacing the one-time anxiety about corporate power is the frequent concern about corporate stasis and incompetence. Some of the effort that corporate managers once directed at exploiting workers and consumers is now committed to gaining, sustaining or advancing their personal corporate position and, quite specifically, their own compensation. Personal profit maximization, that universally acclaimed motivation, can and does extend to those who effectively head the firm.

None of this means that the exercise of political power — the bringing of influence to bear on the state and the public at large — has disappeared. Business firms small and large, individually and collectively as industries, still manifest their economic interest strongly and effectively in the modern polity. But they are now part of a much larger community with political voice and influence that economic development itself has brought into being.

Once, apart from the capitalists, there were only the proletariat, the peasantry and the landlords. These, the landlords apart, were subordinate and silent. Now there are scholars, not excluding students; journalists; television impresarios; professionals of the law and medicine; many others. All lay claim to influence. The voice of the business enterprise is now one among many. Those who would single it out in order to urge the benefits of social ownership are lost in the deeper mists of history. Nor does the experience of the countries for whom public ownership became policy over the last eighty years — the Soviet Union, the Eastern European lands, China — suggest that it enlarges the liberties of the citizen. On the contrary. Accordingly, the principal case for socialism has dissolved. This is recognized. Socialist parties still exist, but none of them is assumed to advocate public ownership in its traditional and comprehensive sense. The British Labour Party's Clause Four, which affirmed support for such a policy and was long seen as a romantic link with the past, has now been formally deleted from the party program.

If socialism can no longer be considered the controlling framework of the good or even the plausible society, neither can capitalism in its classical form. Central is the fact that as the modern economy has developed and expanded, ever more responsibilities have been imposed on the state. There are, first, the services that the private economy does not, by its nature, render and that, with economic advance, create an increasing and increasingly embarrassing discrepancy between the private and the public living standards. Expensively produced television programs are shown to children who attend bad public schools. Houses in the better sections of the city are elegant and clean, while the streets and sidewalks in front are filthy. Books are widely and diversely available in the bookstores but not in the public libraries. Of this, more later.

There is also the wide range of public activities that are necessary for the effective functioning of the private economy. With economic development these also grow in urgency. More commerce requires more highways; more consumption means more waste disposal; for more air travel there must be more airports and more men, women, and sophisticated machinery to ensure the safety of flight.

With higher levels of economic activity, the better protection of the citizen and of the business enterprise also becomes important. Before highways and automobiles there was no need for highway traffic police. As foods have increased in variety, there is increasing consciousness of their nutritional effect — of fats and of being fat. It has become necessary to specify their content, regulate additives and prevent possible contamination. At higher living standards and with greater enjoyment of life, people seek protection as to health and safety from what were once considered and dismissed as the normal hazards of human existence. With economic development, social action and regulation become more important even as socialism in the classical sense becomes irrelevant.

And there is the further fact that the modern economy cannot, without government intervention, ensure a satisfactory and stable overall economic performance. There can be intense and damaging speculation, painful and enduring recession or depression. The appropriate action to control them is much debated, but that it is a responsibility of the state so to do few doubt. Any president or prime minister knows that he or she will be held rigorously and often disastrously accountable at election time for the performance of the economy.

As comprehensive socialism has diminished and disappeared as an acceptable or effective ideology, an opposing, if more limited, doctrine has emerged. This is privatization, the generalized return of public enterprises and functions to private operation and the market economy. As a broad rule, privatization ranks with comprehensive socialism in irrelevance. There is a large area of economic activity in which the market is and should be unchallenged; equally, there is a large range of activities that increases with increasing economic well-being where the services and functions of the state are either necessary or socially superior. Privatization, therefore, is not any better as a controlling guide to public action than is socialism. In both cases the primary service of the doctrine is in providing escape from thought. In the good society there is in these matters one dominant rule: decision must be made on the social and economic merits of the particular case. This is not the age of doctrine; it is the age of practical judgment.

There are, of course, broad tendencies of the modern social and economic system that do bear on public policy and the need for public action. Today's market economy, which so competently supplies consumer goods and services, does so in pursuit of relatively short-run return; that is its measurement of success. It does not invest readily, sometimes not at all, for long-run advantage. Nor does it invest to prevent adverse social effects from its production or its products, which is to say it does not assume responsibility for environmental damage. Of this there will be later mention.

Other examples of public investment beyond the time constraints of the private firm are evident on all sides. The modern jet aircraft is, in substantial measure, the product of military research and development. Much medical discovery has come from publicly supported effort; it could not have occurred within the time and cost limitations to which the private firm or researcher is subject. The most spectacular gain in productivity in modern times has been in agriculture. This was largely the result of state participation — in the United States the work of the publicly supported land-grant college system, the state and federal experiment stations and the publicly supported extension services.

In the years since World War II, economic advance in Japan has been effectively assisted by state-supported research and investment; this has been seen as completely normal. And in all countries the economic system depends on and develops from the state financing of highways, airports, postal services and urban infrastructure of the most diverse and essential sort.

Here then is the lesson. In the good and intelligent society policy and action are not subordinate to ideology, to doctrine. Action must be based on the ruling facts of the specific case. There is something deeply satisfying in the expression of an economic and political faith — "I am firmly committed to the free enterprise system"; "I strongly support the social role of the state" — but this, to repeat, must be seen as an escape from thought into rhetoric.

The matter here urged is especially relevant as this is written. The Republican majority that came to legislative power in the American Congress after the 1994 election was committed to the exceptionally rigorous doctrine formally designated the Contract with America, a present-day equivalent in inspiration if not in content of *The Communist Manifesto*. First came the comprehensive ideological commitment that was aimed primarily against the state, a few favored functions — defense, Social Security, provision for penal institutions, numerous corporate benefits — always excepted. Then followed consideration of the specifics — of the relevance, even urgency, of the public services and functions that were being deleted or curtailed. Now, as of this writing, the retreat from controlling doctrine is under way, allowing for the intervention of practical judgment. This must continue. It is by such means that social decency and compassion, perhaps even democracy itself, are preserved.

4

The Social Foundation

I F PUT in sufficiently general terms, the essence of
the good society can be easily stated. It is that every
member, regardless of gender, race or ethnic ori-
gin, should have access to a rewarding life. Allowance
there must be for undoubted differences in aspiration and
qualification. Individuals differ in physical and mental fa-
cility, commitment and purpose, and from these differences come differences in achievement and in economic
reward. This is accepted.

In the good society, however, achievement may not be
limited by factors that are remediable. There must be economic opportunity for all, a matter the next chapter will
adequately emphasize. And in preparation for life, the
young must have the physical care, the discipline, let no
one doubt, and especially the education that will allow
them to seize and exploit that opportunity. No one, from
accident of birth or economic circumstance, may be denied these things; if they are not available from parent or
family, society must provide effective forms of care and
guidance.

The role of economics in the good society is basic; economic determinism is a relentless force. The economic system in the good society must work well and for everyone. Only then will opportunity match aspiration, either great or small.

Very specifically, the good society must have substantial and reliable economic growth — a substantial and reliable increase in production and employment from year to year. This reflects the needs and desires of a people who seek to enjoy greater economic well-being. In popular discussion and formal economics an improving living standard is an accepted good. More important, socially more urgent, is the fact that such economic performance is essential for the employment opportunity and the income that it offers. To this end, there must be more employment and production, an expanding economy. Economic stagnation cannot be accepted or openly urged as a condition of the good society, although this does, in fact, reflect the quiet preference of many of the better-situated citizens, who prefer it to the risk of inflation or to the stimulative public action that accompanies or ensures steady economic advance.

So long as there is opportunity, there is also social tranquillity; economic stagnation and privation bring with them adverse and widespread social consequences. When people are unemployed, economically deprived and without hope, the most readily available recourse is escape from harsh reality by way of drugs or violence. The practical manifestation is crime and revolt met by futile efforts at repression. The relation of these to deprivation is inescapable. The comfortably affluent parts of the cities and their suburbs are relatively peaceful in the United States, as are those in the other advanced countries. It is on the poverty-ridden streets that the threat or reality of violence exists. This is taken for granted; the only difference of view comes from those, not a few, who blame the disorder on race and ethnic tradition, never on poverty. After the rioting in California in the spring of 1992, the citizens of South Central Los Angeles were held to be given, in some anthropological sense, to antisocial behavior. Not so the superior citizenry of Beverly Hills or Malibu.

The same is true on the larger world scene. It is the poor of Africa and Asia and Central America who slaughter each other; the people of the prosperous lands, on the whole, live peacefully together at home and abroad. It was economic distress in the 1920s and 1930s that helped bring fascism and eventual catastrophe to Italy and Germany. In more recent times, since the fall of Communism, it has been economic hardship and insecurity that have nurtured political conflict and social disorder in the countries of the former Soviet Union.

The lesson for the contemporary American policy and polity is clear. Crime and social convulsion in our great cities are the products of poverty and a perverse class structure, later to be examined, that ignores or disparages the poor. The presently accepted solution is police action, the warehousing of the criminally inclined, an expensive and futile attack on the drug trade. In the longer run and over time, the humane and quite possibly the less expensive solution is to end the poverty that induces to social disorder.

A strong and stable economy and the opportunity it provides are thus central to the good society. There is a further basic requirement. Under the best of circumstances, there are some men and women who cannot or do not participate. In the good society no one can be left outside without income — be assigned to starvation, homelessness, untreated illness or like deprivation. This, the good and affluent economy and polity cannot allow.

For those kept from economic participation by age, the goal of the good society has been long evident and, indeed, is no longer greatly controversial. Ample and secure retirement income for the old is accepted in all the advanced countries. No American politician, however eccentrically motivated, speaks in forthright opposition to Social Security pensions.

But there must also be support for other groups in the society to whom the economy does not provide revenue. The single mother of young children is clearly a case in point. (Possible future starvation for the mother and the young has never been a valid inducement to sexual restraint or birth control.) Similarly the medically or mentally infirm and physically and mentally incapable. And, as now, those who are between jobs and temporarily without income. That all so situated should be accorded basic support must be accepted. Nor should there be anything socially derogatory about being dependent — "being on welfare." Those in need have enough to suffer without being socially stigmatized.

The more obtrusive problem concerns those who, because of neither age, physical disability nor lack of opportunity, choose not to work; of those so disposed there will always be a certain number. They are in conflict, however, with the most frequently cited, socially most compelling of behavioral norms — namely, the work ethic. Nothing is more acclaimed than the latter. Nothing is thought more sharply to define the middle class, normally referred to as the hard-working middle class, as its commitment to the work ethic and, in consequence, its unwillingness to support idleness in the class below.

The avoidance of harsh toil is not, in fact, consistently condemned. In the income structure leisure is not thought socially unacceptable if sought by those in the upper reaches. On the contrary; for the affluent and the rich it has a large measure of approval; it can be a personal and a social virtue. Thorstein Veblen, in his enduring classic, *The Theory of the Leisure Class*, saw well-considered idleness as the prestigious social hallmark of the rich, and so it remains. Intellectuals, not excluding college professors, are known to need recurrent and sometimes extended relief from the pressures of mental toil. We must be tolerant of the preference for leisure as it manifests itself at all levels of our economic system.

In the good society no one can be allowed to starve or be without shelter. The first requirement is that there be ample employment and income opportunity, not enforced inactivity. The major part of the solution thus lies in the overall performance of the economy. Resort to public support must not become necessary because there are no jobs available. But adequate employment being ensured, there must still be a safety net for all. That some will choose not to work must be accepted. Socially compensated idleness unquestionably affronts deep-seated social attitudes; public pressure may, indeed should, be exerted to get able individuals into the work force, the exercise of such pressure being undoubtedly enjoyed by some. Starvation is not, however, a tolerable sanction. Some abuse, as it will be regarded, is inevitable in this part of the welfare system and must be tolerated.

The good society does not seek equality in economic return; that is neither a realizable nor a socially desirable goal. There are those for whom income and wealth and their public manifestation or private contemplation are the ultimate goal and satisfaction; there are others for whom they are not. The Wall Street operative measures the quality of life by his or her income; the poet, actual or aspiring, does not. It is the essence of liberty that these differences in motivation and reward be accepted.

On the other hand, there are sources of income that the good society cannot defend. It is a marked feature of the modern economic system that it provides many opportunities to make money that are either socially indefensible or intrinsically damaging to the economy. Much celebrated in the past was the income from monopolistic exploitation. And more recently the income from larcenous operations in the savings-and-loan industry. And that of the trader on inside information and of the corporate raider and leveraged-buyout operative who leaves a corporation under a heavy burden of debt, with cost to future operations and employment. And, as indicated, the earnings of the corporate executive who, empowered by secure position and controlling fully his board of directors, maximizes his own compensation in accordance with the oldest of motivations in economics. And income from the sale of products that endanger or defraud the consumer or damage the public health or the environment.

The good society must distinguish between enrichment that is socially permissible and benign and that which is at social cost. The energy and initiative of the individual who devotes himself to the often barren goal of personal enrichment can be serviceable to the economy. They can also result in methods that are socially damaging. The good society must therefore assume the essential, difficult and intensely controversial task of making and making effective the distinction. Informed pursuit of gain is benign. Use of insider information or the conveying of false information is not. The past experience with highly leveraged corporate takeovers and, in the main, with junk bonds has been bad, and there has been heavy damage from various alluring investment designs. There are, as I have noted, no universal rules in these matters — nothing can be decided by a recourse to free-market, capitalist, liberal or socialist principles. Here again the world of intelligent thought and action, not of adherence to controlling doctrine.

There is also as regards income the matter of what may be called the social claim. In the good society there is, must be, a large role for the state, and especially on behalf of the less fortunate of the community. This need must be met and paid for in accordance with ability to pay. Basic justice and social utility are here involved. A loss of income at the margin is less painful for the rich than for the less affluent. It also contributes to the efficient functioning of the economy. The poor and those of average income spend reliably from what they earn; the rich do not. Thus, progressive taxation has a stabilizing role in helping to ensure that what is received as income is returned to the market as demand for goods produced. It has been suggested that the attempt to maintain after-tax income serves to induce the rich to economically productive effort; that, however, may be carrying the case too far.

While the economy must accord everyone a chance both to participate and to advance according to ability and ambition, there are two further requirements. There must be a reasonable stability in economic performance; the economic system cannot recurrently deny employment and aspiration because of recession and depression. And it may not frustrate the efforts of those who plan diligently and intelligently for old age and retirement or for illness or unanticipated need. The threat in this case is, of course, inflation — the diminished purchasing power of money — and with it the loss of provision for the future.

Future security in life is based normally on the assumption of stable or reasonably stable prices. There are some who have the protection of indexing, income that rises along with prices; many do not. The good society therefore must honor the expectation of reasonable price stability. This cannot, in a well-functioning economy, be absolute; some price inflation is inevitable. It must, however, be within close and predictable limits. The good society does not accept what John Maynard Keynes called the euthanasia of the rentier class.

Finally, the good society must have a strong international dimension. The state must live in peaceful and mutually rewarding association with its trading partners on the planet. It must be a force for world peace; it must work cooperatively with other nation-states to this end. War is the most unforgiving of human tragedies, and is comprehensively so in an age of nuclear weaponry. There must also be recognition of, and effective support for, the needs and hopes of the less fortunate members of the world community.

Such are the broad specifications of the good society in its social dimension. Employment and an upward chance for all. Reliable economic growth to sustain such employment. Education and, to the greatest extent possible, the family support and discipline that serve future participation and reward. Freedom from social disorder at home and abroad. A safety net for those who cannot or do not make it. The opportunity to achieve in accordance with ability and ambition. A ban on forms of financial enrichment that are at cost to others. No frustration of plans for future support and well-being because of inflation. A cooperative and compassionate foreign dimension.

The specifications are rather evident, even commonplace, and, with some very notable exceptions, acceptable, especially in the oratory of the time. It is the actions necessary to achieve these ends that are more controversial.

5

The Good Economy

THERE IS no serious doubt as to the economic basis
of the good society. As sufficiently noted, there
must be employment opportunity for all willing
members. This means, with a growing population and
greater aspiration, a steady expansion of the economy and
therewith a steady and reliable increase in the number of
workers employed.

The central problem with which the good society must
here contend is the painful tendency of the modern economy to periods, sometimes prolonged, of recession and
stagnation, accompanied, inevitably, by more unemployment. These recurrent episodes, not continuous vigorous
expansion, are a basic feature of the market system. So in
modern times is continuing unemployment, even in periods of marked growth and well-being.

Few matters have been subject to more relentless study
than economic fluctuations, what anciently were called
the business cycle. The essence of the phenomenon is not
at all obscure. The modern economy requires a steady
flow of purchasing power — in economic terms, aggre-

33

gate demand — that is sufficient to utilize the available productive capacity, encourage the requisite expansion therein and employ all available workers. Once extensively identified with industrial factory wage work, the working force thus employed now extends to a broad range of occupations — services, the arts, entertainment, the higher levels of technology, education, much more. While some difficulty in fitting specialization to need is inevitable, it is the total employed in the labor force that, at any given time, sets the upper limit on the productive capacity of the economy. The flow of aggregate demand for goods and services must keep the economy at or near that limit. The failure to do so — resulting in cyclical or enduring unemployment — is inconsistent with the goals of the good society.

The cyclical tendency of the basic economic system must be assumed. There are diverse causes, but the evident and reliably persistent one is speculative excess in good times, which results eventually in the curtailment of investment and consumer expenditure; this reduction in the flow of aggregate demand has inevitable consequences for the production and employment thus sustained. The speculative episode in one form or another — in securities, real estate, junk bonds, the mergers-and-acquisitions mania of the 1980s with its depressive debt accumulation, all in descent from the frenzy for tulips in Holland in the seventeenth century and the great South Sea Bubble in the eighteenth — is, to repeat, an inescapable feature of the system. It has been experienced in the United States since the earliest days of the Union. Recession or depression and unemployment inevitably follow. Better understanding, appropriate regulation, greater common sense, can perhaps help to control the boom, but, as a practical matter, attention must be concentrated on mitigating the distress and hardship and especially the unemployment that result from this basic cyclical instability in economic life.

The stabilization of the flow of aggregate demand is the vital factor. Aggregate demand has three decisive components: consumer expenditure, expenditure for private investment and expenditure from the fiscal operations of the state — from government spending that exceeds or falls short of tax receipts. On occasion, the obvious ends economic contention.

If the flow of purchasing power — of aggregate demand — is insufficient to sustain a high level of economic activity and growth, it is commonly believed that certain readily available and greatly benign measures will restore consumer and business confidence; this will lead, in turn, to increased private investment expenditure and more consumer spending. The measures so favored are oratory and public and private prayer, entreaties that the federal budget be balanced and suggestions that public regulatory action be restrained. These have not been shown by past experience to be greatly effective. Nor are they widely so regarded except by those resorting thereto.

With time, recessions or depressions — there is no formal difference between the two, although some economists have sought to distinguish between them — do come to an end. The excesses and inevitable losses that were caused by speculation and are the invariable mark of good times retreat in memory and effect. Spending and investment revive. The waiting period, however, is painful, and especially to those who are the most vulnerable in the economic world. The good society, accordingly, must have an effective design for countering these periods of distress and ensuring a steady, reliable increase in production and employment.

There are three substantive lines of corrective action that will accomplish this, will increase the flow of aggregate demand as required. Taxes can be lowered, thus releasing to the consuming public more revenue to be expended on private consumption. In a view much favored by those whose tax payments are thus reduced, one that has already been suggested, this is also held to encourage personal and business initiative and investment, resulting in a further addition to aggregate demand. No one is believed to be so inclined to desuetude, so aroused by the prospect of more income, as the affluent taxpayer.

Second, interest rates can be reduced by central-bank action, thus encouraging business and consumer borrowing and investment or expenditure, which add to the flow of aggregate demand.

Further and finally, the government can contribute directly to the flow of demand by new expenditure in excess of tax receipts — by a deliberately accepted or deliberately increased deficit. By one or a combination of these steps aggregate demand can, or so it is held or hoped, be kept at a level that will cause business and the government to reach out for all available workers.

There is, unfortunately, a wide difference in the effectiveness of these several public actions. There is also the problem of inflation. And there is the already-observed socially concealed preference of many in the society for stagnation and unemployment over the measures that contribute to, or ensure, economic growth and high employment.

Action on interest rates, commonly referred to as monetary policy, has the highest establishment approval as an effective measure against stagnation and unemployment; it must, accordingly, be the first for consideration. The requisite authority is available to, and assumed by, a central bank; in the United States, the Federal Reserve. It is considered the special virtue of monetary policy that it is removed from the pressures of the democratic process; the necessary action is taken by those responsible, in socially hygienic detachment from ordinary political contention, influence or control.

As regards monetary action, there is also the peculiar magic that is thought to be at the command of those intimately involved with financial matters — a magic that is beyond the comprehension of even the most informed layman. The point must be emphasized: nothing in modern attitudes is believed more to signify exceptional intelligence than association with large pools of money. Only immediate experience with those so situated denies the myth.

The serious flaw in monetary policy is that it may have little or no effect on the flow of aggregate demand. As noted, the lowered interest rates are assumed to work against depressive conditions by encouraging consumer borrowing and expenditure and business investment. The latter responds to the lower cost of borrowing and therewith the improved possibility for profit. But when times are poor and unemployment is high, lower interest rates do not reliably inspire consumer expenditure; depressive attitudes, including those which are the product of unemployment or uncertain employment, are in control. And at such times, excess business capacity being evident, business firms, old and new, may not be encouraged by low interest rates to borrow and invest and so contribute to the income flow; the larger prospect is too uncertain. There is also the adverse effect of low rates on those whose income comes from interest — a reduction in their contribution to aggregate demand. None of this, however, discourages faith in monetary action as a decisive economic instrument. Quasi-religious conviction here triumphs over conflicting experience.

Tax reduction is also celebrated as a way to sustain aggregate demand during recession. With lowered taxes on income or consumption the individual will have more funds at his disposal to add to the flow of demand. Also, inspired by the prospect of more income after taxes, producer energy and investment will be enhanced.

Here again the hope is at odds with the reality; there is no certainty that the funds released by tax reduction will be invested or spent. In bad times people and firms so benefited may well choose to hold on to their money. Also, a disturbing part of the support for tax reduction as an antidote to economic stagnation and unemployment comes from those whose tax burden would thus be eased. From personal advantage, it is said, will come public gain. The opposite effect of taxation in the upper-income brackets is, in fact, true: there, taxes may compel the expenditure of funds that would otherwise go unspent or uninvested. Tax reduction can serve only what John Maynard Keynes called, in a noted phrase, liquidity preference. The desire to hold cash or its equivalent does not add effectively to aggregate demand.

As a way to stimulate demand in times of negative growth or stagnation, there remains only direct and active intervention by the state to create employment. In an ideal world this last would not be necessary. In the real world of recurrent and long-continued stagnation there can be no effective alternative.

The specifics of the correct public measures against recession and depression are clear. Interest rates should, indeed, be reduced, for whatever effect this may have. But the only truly substantive action is for the government to move to provide jobs for those for whom unemployment is otherwise inevitable. In doing this, it must borrow and accept the reality of a larger deficit in the public accounts. The deficit, as will presently be noted, must not be seen as a barrier to effective public action, for by stimulating economic activity it increases earnings and tax receipts. Improvements to the public infrastructure — roads, schools, airports, housing — that are made by those newly employed also add to public wealth and income. Public borrowing can, over time, be a fiscally conservative act.

When the economy recovers and public revenues rise, there must then be the discipline that brings stimulative expenditure to an end. Taxes must be kept at previous levels or increased as a counter to speculative excess and, at the extreme, the inflationary pressure of demand on markets.

There is nothing easy about this broad course of action. An influential body of opinion now dismisses it as beyond the collective intelligence of the modern polity. Once again the unfortunate fact asserts itself: there is no effective alternative. What is dismissed as functionally difficult, ideologically passé, is the only way to prevent recurrent periods of stagnation and unemployment.

In the winter of 1995, the newly elected Republican majority in the United States Congress, with some Democratic support, came within a vote or two of submitting to the states a constitutional amendment calling for a balanced budget in all but wartime. This could have been the economically most regressive legislative proposal of recent years, a hard competition to win. It would have required tax increases and government expenditure reductions when the ordinary flow of government revenues was already reduced by recession or depression. And it would have allowed more public spending and tax reduction when times were good, adding thereby to the general speculative and inflationary mood. There could have been no better design for enhancing economic instability.

This regression was defeated, though only by the minority required to block congressional passage of a constitutional amendment. It was further evidence that the necessary action to counterbalance boom and recession, the inescapable feature of the market economy, is far from accepted. Compelling attitudes still seek not the good and stable economy but the painfully unstable one.

A final point must be mentioned. Economic failure, unemployment, is regularly blamed on the workers. A standard response to higher unemployment figures is the call for better worker training. That is the politically respectable remedy. Education, training, is, indeed, central to the good society, a matter elsewhere emphasized. But it is not relevant to the cyclical downturns that are here discussed. When depression or recession comes, both the trained and the untrained, the educated and the ignorant, are affected. Of this there should be no doubt. A call for better-prepared workers as the remedy for recession-induced unemployment is the last resort of the vacant liberal mind.

The good society must contend with the depression, recession or stagnation that afflicts the modern market economy. But it must also contend with the problems of high and full employment, inflation and the deeply concealed preference of some, indeed many, for economic stagnation. And there is the widely controversial issue of the deficit. These are the subjects of the next chapters.

6

Inflation

A s earlier indicated, there are two especially difficult problems that intrude on the search for the good economy. One is the likelihood, indeed almost the certainty, that at full or near-full employment, and with a rewarding rate of economic expansion, there will be some inflation. And there is the further fact that a large and influential sector of the modern polity has no personal quarrel with stagnation and unemployment, preferring them greatly to the measures that would effectively address them or to the risk of inflation so invited. Only the fact of this preference must not be admitted, not even mentioned. To be publicly for recession or stagnation would be politically less acceptable than a vigorous stand in support of sexual harassment.

With an approach to full employment — a job opportunity for all willing workers, enhanced income for others — the threat and the reality of inflation become, along with the deficit in the public accounts that may have stimulated such recovery, the most discussed of all economic questions. And the controlling causes of inflation are not in doubt. The flow of demand that clears markets, expands production, enhances the need for workers, also allows ineluctably for price increases. This opportunity will then be exploited or compelled. And where there is general employment opportunity, there will always be general, regional or specialized job scarcity. Such scarcity will be overcome by the offer of higher wages in the knowledge that in the strong market that has led to the wage increase, the added costs can be passed on to the consumer. And the higher wages will lead to more demand, more pressure on markets.

The strong market and favorable return to the employer also, needless to say, stimulate the claims of the trade unions, in the United States a declining but still appreciable and socially essential force. These claims, in turn, produce both the justification and the necessity for higher prices.

Economics does not always celebrate its insolubles. It does so, however, in the matter of employment and inflation. Many years ago the Phillips Curve of A. W. Phillips, of the London School of Economics and the Australian National University, identified the clear choice — the trade-off — between high employment and inflation as against unemployment and relatively stable prices. This trade-off is present in all accepted thought.

In recent times, however, there has been a significant shift in the preference as between inflation and unemployment. Once unemployment was the controlling fear; full employment was the prime test of economic performance. So it remains in much of the reputable economic expression. But the deeper reality is that inflation is now considered by the most influential part of the modern polity to be the central threat to good economic performance; stable prices are the dominant objective. Unemployment, in this view, has become a price-stabilizing instrument. This reflects a new reality, not often so bluntly described but visibly, even obtrusively, present.

The controlling fact is that in the modern economy and polity those who have political voice and influence are more damaged by inflation than by unemployment. Unemployment is suffered by those afflicted and by their families; their pain can readily be tolerated by those who do not experience it.

Unemployment has, in fact, some socially and economically attractive effects: services are well staffed by eager workers forced thereto by the lack of other job opportunity; employed workers, fearing unemployment, may well be more cooperative, even docile, as may their unions. And, even more significantly, for most citizens, including those with influential political voice, joblessness is not a threat.

Inflation, in contrast, spreads its net widely in the modern economy. The many who live on fixed incomes, on pensions, on accumulated savings, fear it as they do not fear unemployment. Even if income return is indexed to rising living costs, a sense of insecurity is still instilled by higher prices. The increases are seen every day; the indexed adjustment comes only at intervals of as long as a year. Price stability seems better by far.

Prominent among those preferring price stability over unemployment is the financial community. This includes central banks in which, in the case of the Federal Reserve System, the bankers are accorded a statutory voice. And commercial banks, investment firms and the larger financial world. All who lend money wish to have it returned with more or less equivalent purchasing power. This wish, inflation directly invades. And there is a powerful subjective effect here as well. The avoidance of inflation has independent standing as a decisive test of the quality of financial management. Inflation casts a long shadow over that management, showing that it is defective. A competent central banker is one who minimizes inflation. He — this is not a domain of women — is subject to no similar test as regards unemployment.

While on occasion this is openly expressed — "Unemployment is not my business" — central bankers and the financial community in general treat the price-stabilizing role of unemployment with a certain reticence. It is accepted that a too great or too rapid reduction in unemployment is to be feared and regretted; what is avoided is openly expressed approval of a large supply of eager job-seekers and the price-stabilizing function they perform.

The good and achievable society cannot hope to reconcile full employment with fully stable prices. It can, however, do something to minimize the conflict between the two. Even in a world of diminished union power, there can be recognition of, and restraint on, the wage-price spiral. Wage settlements can be held within the framework of existing price structures. This has long been an accepted feature of trade union bargaining in Europe; not surprisingly, it has come to be called "the European model." On its part, management must show its respect for worker restraint by keeping its prices stable.

In past times governments of both conservative and liberal disposition, including that of the United States, have urged wage and price restraint and, from time to time, enforced it by wage and price controls. This, however, is in conflict with the basic structural character of the market system and likewise with powerful economic and public attitudes and beliefs. The most that can now be urged is a sense of responsibility on wage-price negotiation that reflects the larger public interest.

The choice between unemployment and inflation cannot be evaded; it must be faced. The good society cannot relegate some parts of its population to idleness, social distress and economic deprivation in order to achieve price stability. As necessary, the lesser evil of price increases must be accepted. There is never a case for serious inflation — for any sharp deterioration in the purchasing power of money — but the progressive economic expansion that draws the most workers effectively into employment will, inevitably, mean some upward movement in prices. So, without damaging effect, it has in the past. So also it will in the future. This is socially better than stability achieved by the depressive effect of widespread enforced leisure.

It must be understood that there cannot be full employment and stable prices at the same time. The good society, accordingly, must move to offset or minimize the unpleasant effects of both. Unemployment compensation must be generous; episodic abuse cannot be a case against it, nor can acceptance of it be, in any sense, socially derogatory. It is an essential and important part of the system.

Similarly, as it is recognized that a moderate rate of inflation is inevitable, there must also be action to mitigate its negative effect. Specifically, there must be general indexation of fixed and contractual income, of pensions, the wages of teachers and civil servants, and of the level of the basic safety net. Also and, as this is written, most important, of the minimum wage. Indexation has become a widely established practice; it must be extended as necessary and regarded as normal. The setting of interest rates must also reflect the expectation of increasing prices. This, however, is a matter on which the financial community is now more than adequately sensitive.

As noted, much of the adverse attitude toward inflation comes from the rentier class. This is a large community in the modern society, and its deep opposition to inflation and the strength of its expression cannot be doubted. Both come from its clear preference for high interest return combined with stable prices.

There is the further fact that lurking in the financial mind and in much of public opinion is what has been called the pregnancy theory of inflation. As a woman cannot be a little pregnant, so there cannot be a little inflation; something more is inevitable. This is clearly nonsense; inflation can increase or diminish according to the motivating forces. This has been the experience of past years, even decades. Prices have consistently edged up, and without eventual disaster or, as commonly described, hyperinflation.

In the years following World War II, there was high employment and vigorous economic growth, along with a modest annual increase in prices. The latter did not signify that things were out of control. What was different then was that those subject to insecurity in employment were not singled out as the solution to the threat of inflation.

In delineating the characteristics of the good society it would be agreeable to specify that full employment and stable prices should be achieved simultaneously. This, indeed, has been cited as a goal in much past comment. Here, as always, the utopian in conflict with the achievable. A low level of unemployment is necessary; it is a goal that cannot be compromised. But combining this with absolute price stability is not within the range of the possible.

7

The Deficit

THERE ARE TIMES in modern history and experience when the enunciation of even the most elementary common sense has an aspect of eccentricity, irrationality, even mild insanity. Such is the risk that is run by anyone in the United States today who challenges the current commitment to reducing and eliminating the government deficit, this being the overall excess of expenditure over income in the public accounts. As just noted, a requirement for a balanced budget recently came within a vote or two, if the states subsequently agreed, of becoming a constitutional mandate, not less as to depth of legal commitment than freedom of speech or the rights of private property. That a given expenditure could increase the deficit has become a decisive objection to it, and this is so even when the most needful of purposes, involving the most needful of citizens, is under discussion — in substantial measure in recent times, especially then. In the American experience certain government expenditures remain outside the public anxiety about the deficit, those for the military, to which I will later come, being the impressive case. Those for the poor most definitely do not.

What in the good society is the controlling rule as regards public borrowing and the creation of the public deficit?

There is no law or tradition, I will argue, that requires a balanced budget — equal revenues and expenditures on an annual basis. This does not mean, however, that the budget and budget deficits in the modern state can be treated casually; a high measure of intelligence and discretion is always required. What is essential is that the intelligence and discretion be exercised within the relevant framework. Let me be specific, an exceptional tendency on this particular matter.

There are three broad categories of public expenditure. There are those which serve no visible present or future purpose; there are those which protect or enhance the current economic or other social condition; and there are those which bring or allow of an increase in future income, production and general well-being.

First as to expenditures with no good or necessary purpose. It must be accepted that no institution is perfect, and certainly not the modern state. There is the tendency in any great organization, public or private, to an excess of personnel — to the universal desire of all in the organizational hierarchy to employ additional subordinate talent or what is so denoted. Life is always enhanced when one has others to do one's work and one's thinking. The number so employed is a common measure of the position and prestige of their employer, while adding to the total expenditure of the employing organization. There is also expenditure that responds to political or economic interest, not to the larger public need or desire. And there is expenditure that survives the purpose for which it was originally intended and which it once served.

To the extent that purposeless expenditure escapes control and elimination — one of the prime tasks of public administration — it must be a current charge against current revenue. No one, the recipients of the particular largesse always excepted, can argue that such expenditure should be covered by public borrowing. That subsequent generations should be made accountable for present waste has no public supporters. What is troubling in this instance, however, is not the principle but the practical fact that the waste involved is not easily detected. That is a task which is enormously complicated by the long-established tendency to describe as waste what, in fact, beneficially affects some other and needier part of the public community.

The second and very large category of public expenditure that must be covered by tax revenue is that for the current, everyday operations of the government — for those functions which are urgent today but have no clearly recognizable time dimension. These include the vast range of government activities — the enforcement of law, the routine conduct of foreign policy, governmental support to industry and agriculture, peacetime support to the military (which, as noted, is subject to peculiar considerations later to be examined) and much else. There is no economic or social justification for borrowing for these tasks, thus adding interest charges to the eventual total cost. Subject to the larger fiscal policy effects that defend against recession or depression, the current democratically decided functions of the state should be paid for by taxes and other current revenues.

There remain those government expenditures which are intended to improve future well-being and economic growth or which so serve. Here, borrowing is not only legitimate but socially and economically desirable. Similar borrowing in the private sector of the economy is both accepted and wholly approved even by the most eloquent, frequently vehement, opponents of the public deficit.

Specifically, what the modern business enterprise takes for granted is also appropriate policy for the state. Expenditure for current production should be a charge against current revenues; investment that enhances future income and advantage need not be. Here, borrowing should be accepted, normal. Interest and amortization costs should be charged against revenues; capital expenditure should not. Broadly speaking, this is now the rule as regards local government in the United States. Only the federal government denotes all capital expenditure as current expense. No logic defends this accounting, only convenience, tradition and political rhetoric and error.

Where public expenditure promotes or, indeed, is essential for the future growth of the economy — increased production, employment and income from which to sustain future public revenues — borrowing is fully acceptable. This cannot be considered loading costs on future generations, for they will be the beneficiaries and it is appropriate that they pay. Assuming that tax rates remain generally the same and the economy is otherwise stable, such payment will come out of the expanding future revenues. These expanding revenues are, in greater or lesser measure, the product of the longer-term investment. It is to facilitate such gains that, in the largest sense, they are made.

There is much government expenditure that results in this future reward. That for public works is the obvious and, indeed, most commonly mentioned example. Investment for improved transportation facilities, not excluding air traffic and its future control, is equally obvious. Later generations should pay for what they will use. The case is the same with borrowing for the improvement of public postal services; such borrowing is something the competing private postal services take for granted. But the obvious is not the total or even a dominant part of the argument.

Investment in health care means a more productive work force in the future: because of good health, there will be less need for medical expenditures. Similarly investment in programs to restrain drug, alcohol and tobacco abuse. Children protected and rescued from poverty by

54

welfare will become productive citizens. Such current out-lay will nourish future productivity and yield the additional income that, when taxed, will pay the interest and amortization of the increased debt.

These are, however, only the glimmerings of the larger picture. Nothing will so improve future income and output — the yield of the economy in general — as the educational qualifications of the people. It follows that when investment for the future is considered, nothing will so effectively underwrite future return as that in education — in the improved intelligence and productivity of human beings. On the other hand, much educational as also medical and like expenditure should be considered a current charge; there is no longer-range investment effect. The problem arises in distinguishing between the two.

There is, in fact, no way by which expenditure for the current needs of education, health, basic welfare or for many other public services can be separated from that which will augment future income. And it may well be doubted whether an effort so to identify expenditures for future wealth creation is worthwhile. There is no possibility of a numerical as opposed to a conceptual distinction. One broad rule, however, does exist.

Always assuming overall efficiency in public administration and intelligence in the assessment of public functions, the deficit and the resulting interest charges should increase over time in a constant relationship to growth in the economy as a whole. If they increase more rapidly, it must be asked whether they include some expenditures that are not contributing as expected to economic growth. If they fall behind, there is at least a question as to whether there is adequate public investment for needed economic expansion. Any exact calculation as between current and capital components of the budget being impossible, we must fall back on aggregates.

The cost of sustaining the public debt should be roughly in keeping with the increase in the means available to pay for it. In specific economic terms, interest charges on the debt should be a fairly constant percentage of the growth in aggregate revenues from which it is paid. In the United States, the concern for the public debt notwithstanding, this relationship has been broadly stable in recent times. In the 1980s, there was a sharp increase in federal interest charges as a percentage of Gross Domestic Product, that is, of the larger ability to pay, in consequence of the Reagan military buildup and the exceptionally casual or, as some have described them, ultra-Keynesian budget policies of that time. Thereafter interest costs as a percentage of GDP showed a modest decline and, at this writing, remain basically constant. Future generations will indeed be charged with paying for some current public expenditure. But subject to the qualifications previously noted, they will be doing so out of the greater income and well-being that the expenditure has helped to bring about.

Two matters remain. There is, first, the need to move on from a rigid annualized view of the budget process. When times are good and government revenues are strong, the deficit should be reduced. In effect, there should be a larger charge for investment for future economic growth. And by the same token, when there is recession and associated distress and unemployment, public capital investment and employment should increase, as will, inevitably, the deficit. The case for this, a compensatory fiscal policy, has already been made.

The basic economic policy of the good society is public expenditure in step with future economic growth and well-being. It is thus in accord with the means for interest payment and amortization, subject to the necessary adjustment to the current economic condition of prosperity or recession. It will be argued that this is unduly demanding as to public action. Democratic government operates according to less sophisticated, more elementary rules. Such could be the case. But no one should suppose that the guidance of the modern economy is a simple matter. Perhaps, indeed, we will fall short of the ideal in such public management. With the current emphasis or, in any case, the current oratory on deficit reduction, this is what is happening in the United States and also in other industrial countries. There can be no excuse, however, for not knowing what the right policy — the fiscal basis of the good society — must be.

The budget deficit is now being used, it has been noted, as an instrument against socially necessary but politically resisted public policy. Interposed against the widest range of social action is the argument that it would increase the deficit and the tax load on our grandchildren. This, it will be evident, is errant and self-serving foolishness. Properly viewed, the deficit can be a source of support and benefit to future generations — an enlargement of their general prosperity and ability to pay. So it has been in the past; so it should be in the future.

There is yet further opposition to accepting a deficit and public capital formation. To do so, it is held, depletes the savings supply and robs the private sector of needed investment funds. That argument is advanced when interest rates are low and an abundance of funds are seeking investment. And similarly when rates are high and central-bank action could make them lower and presumptively encourage investment. The argument opposes private investment for however frivolous the consumer product or service against public investment of whatever social urgency. The savings-conservation case should be viewed with amused detachment as part of the general use of the deficit as an instrument against enlightened public purpose.

Budget and fiscal policy are the most demanding of modern economic policies, and especially as they concern the deficit; they are the foundation on which much other policy depends. No one should minimize the problems that are involved. Or the initiative and the restraint that are required. Here, the economic basis of the good society comes to its sharpest focus.

8

The Distribution of Income and Power

THE GOOD SOCIETY does not seek equality in the distribution of income. Equality is not consistent with either human nature or the character and motivation of the modern economic system. As all know, people differ radically in their commitment to making money and also in their competence in doing so. And some of the energy and initiative on which the modern economy depends comes not only from the desire for money but also from the urge to excel in its acquisition. This last is a test of social excellence, a major source of public prestige.

A strong current of social expression and thought has held that there is, or could be, a higher level of motivation if there were an egalitarian level of reward — "From each according to his abilities, to each according to his needs." This hope, one that spread far beyond Marx, has been shown by both history and human experience to be irrelevant. For better or for worse, human beings do not rise to such heights. Generations of socialists and socially oriented leaders have learned this to their disappointment and more often to their sorrow. The basic fact is clear: the good society must accept men and women as they are. However, this does not lessen the need for a clear view of the forces controlling the distribution of income and of the factors forming attitudes thereon. And of how, in a wholly practical way, policy on income distribution should be framed.

There is, first, the inescapable fact that the modern market economy (in the now-approved terminology) accords wealth and distributes income in a highly unequal, socially adverse and also functionally damaging fashion. In the United States, now the extreme case among the major industrial countries, the Federal Reserve, an impeccable source said, as reported in the *New York Times*, that the top 1 percent of American households owned nearly 40 percent of the nation's wealth in 1989, the top 20 percent more than 80 percent. The lowest-earning 20 percent of Americans had 5.7 percent of all after-tax income; the best-rewarded 20 percent had 55 percent. By 1992, the top 5 percent were getting an estimated 18 percent, a share that in more recent years has become substantially larger, as that of those in the poorest brackets has been diminishing. This, the good society cannot accept. Nor can it accept intellectually the justification, more precisely the contrivance, that defends this inequality. The latter is one of the most assiduously cultivated exercises in economic thought. Never, however, does it quite conceal the fact that the economic and social doctrine involved is subordinate to the pecuniary purpose (and forthright greed) that it serves.

Specifically, it is held that there is a moral entitlement: the man or woman in question has the right to receive what he or she earns or, more precisely, what he or she receives. This can be asserted with emphasis, on occasion with asperity and often with righteous indignation. It encounters opposition, however, in both history and hard fact.

Much income and wealth comes with slight or no social justification, little or no economic service on the part of the recipient. Inheritance is an obvious case. So also the endowments, accidents and perversions of the financial world. And the rewards that, from its personal empowerment, modern corporate management bestows on itself. As noted, the modern corporate management is committed, as in all orthodox economic doctrine, to profit maximization. Because it is exempt to a substantial degree from stockholder control or restraint, it extensively maximizes return on its own behalf. With compliant boards of directors of its own selection it effectively sets its own salary, provides its own stock options, establishes its own golden parachutes. That such return is unrelated in any plausible fashion to social or economic function is largely accepted. The frequent and sometimes fervid assertion of such function is a cover story for the patently improbable.

The rich have a certain reluctance in defending their

wealth and income as a social, moral or divine right, so their only possible resort is the functional justification. From the undisturbed and admittedly unequal distribution of income comes the incentive to effort and innovation that is in the service of all. And from the income so distributed come the saving and investment that are for the advantage of all. The rich and the affluent do not speak in defense of their own good fortune; they speak as the benign servitors of the common good. Some may even be embarrassed as to their worldly reward, but they suffer it, nonetheless, as a service to the general well-being. Social and economic purpose is adjusted to personal comfort and convenience. This, all in the good society will recognize.

Further, there is the protection that the peculiar class structure of the United States accords the affluent and the rich. All reputable reference concerning class structure emphasizes the middle class. There is an upper and a lower class, but these are back in the shadows. Although it is rarely so designated, for practical purposes we have a three-class system consisting of only one class, an arithmetic novelty. And the middle class, which is so dominant, then provides protective cover for the rich. Tax reduction on behalf of the middle class extends on to the very affluent. The upper class does not, in such discussion and action, separately exist. Such is the political attitude. There is a strong functional effect here as regards the working of the economy.

As to the income going to the very rich, there is, to repeat, the exercise of, in the economist's term, liquidity preference — the choice between consumer expenditure and investment in real capital or merely holding the money in one or another form of effective idleness. This is a choice as to the use of income that the individual or family of modest means cannot exercise. They are under the pressure of more urgent need; that they will spend the money they receive is thus certain. Accordingly, income that is widely distributed is economically serviceable, for it helps to ensure a steady flow of aggregate demand. There is a strong chance that the more unequal the distribution of income, the more dysfunctional it becomes.

What, then, is the right course as regards the distribution of income? There can be no fixed rule, no acceptable multiple as between what is received by the rich and what goes to the poor. Or, indeed, as between what is earned by management at the top and what is earned on the shop floor. The basic character of the system is here involved. It does not lend itself to arbitrary rules. What is necessary are strong ameliorating actions that reflect and address the inherent and damaging inequality.

There is, first, the support system for the poor. The attack on inequality begins with a better break for those at the bottom. This has already been stressed.

There is, second, as also discussed, the need to deal with the dominant tendencies of the financial world. Insider trading, false information in investment promotion, deviant investment behavior as in the case of the great savings-and-loan disaster, the corporate takeovers and the recurrent episodes of speculative insanity, all unfavorably affect income distribution. Measures that ensure elementary honesty in financial transactions and a better understanding of the speculative episode have a useful leveling effect.

There is, third, the need for stockholder and informed public criticism to address the personal income maximization of corporate management. Independent of such stockholder and public restraint, the corporate managerial take has, as already noted, become a major cause of socially adverse income distribution. The only answer here is united action by the stockholder owners who are thus disadvantaged. The chance for such action is, it must be conceded, not great. Those who own the modern corporation are notably passive as regards their personal exploitation.

There remain two lines of affirmative public action looking toward a more equitable income distribution, one of which is of decisive importance.

The first is for the government to remove the present tax and expenditure concessions to the affluent. In recent times these have achieved a measure of recognition under the cognomen of corporate welfare. Included here are diverse business subsidies and tax breaks; support to agricultural producers who are already in the higher income brackets, especially a lavishly endowed sugar monopoly and subsidies to tobacco production; export subsidies, including those for arms exports; and, bulking largest of all, the vast payments to the now recidivist weapons producers, of which more later.

However, the most effective instrument for achieving a greater measure of income equality remains the progressive income tax. This has the central role in accomplishing a reasonable, even civilized, distribution of income. Nothing else, it may be added, is subject to such highly motivated and wholly predictable attack. The good society, on the other hand, affirms its purpose; it also assumes that there will be strong, articulate, even eloquent resistance from those so taxed. They will especially allege the deleterious effect of the tax on incentives. As earlier suggested, it could be claimed with equal improbability that a strongly progressive income tax causes the rich and the affluent to work harder, more imaginatively, in order to sustain their after-tax income. Referring to past experience, it can, indeed, be pointed out that the American economy had one of its highest rates of growth, its highest levels of employment and in some years a substantial budget surplus in the period immediately following World War II, when the marginal rates on the personal income tax were at a record level.

The basic need, however, is to accept the principle that a more equitable distribution of income must be a fundamental tenet of modern public policy in the good society, and to this end progressive taxation is central.

The distribution of income in the modern economy derives ultimately from the distribution of power. This, in turn, is both a cause and a consequence of the way income is shared. Power serves the acquisition of income; income accords power over the pecuniary reward of others. The good society recognizes and seeks to respond to this traditionally closed circle.

Its response is the empowerment and public protection of the powerless. In the market economy the natural focus of power is the employer, most often the business firm. The right of workers to join together and assert a countervailing authority must be central and accepted. As those who organize to invest enjoy the protection that the state accords the corporate structure, so those who organize to enhance income or improve working conditions should have a broadly equivalent protection for their organization.

In modern times, especially in the United States, the empowerment of workers has been diminishing in its general effect. Trade union membership as a proportion of all workers has sharply declined, partly in consequence of the decline in mass-production, mass-employment industry, partly because of the aged lethargy of the trade union movement itself. The good society seeks, where possible, to reverse this decline in trade union power, for worker organization remains a major civilizing factor in modern economic life.

For many workers, however, organization is not now a practical solution. This is especially true in the widely dispersed service industries. As was once the case with the employment of women and children, direct action by the state on behalf of those in need outside the unions is required, including provision for health insurance and unemployment compensation and, currently most important, a socially adequate minimum wage. In the good society the last is an absolute essential. That it will diminish employment opportunity, the argument most commonly made in opposition, may be dismissed out of hand, for that is, invariably, the special plea of those who do not wish to pay the wage, and it is without any empirical support. (Even were it at cost to the employment of the few, it would still be justified as the protection of the many.) Along with a basic safety net the good society must also protect the working income of its least favored members.

9

The Decisive Role
of Education

THERE ARE few subjects that are more intensely discussed than the role of education in the modern society, and particularly its connection with economic purpose. Any and all analysis of the competitive position of the American economy focuses on the importance of a well-educated, occupationally qualified labor force. The point is further emphasized in the frequent references to expenditure for education as human investment. Investment is traditionally for enhanced economic return; education is thus an aspect, more precisely a component, of larger economic policy. This is a belief that calls for examination in the good society.

That education does serve economic purpose is not in doubt. This has long been recognized. In the last century in the United States education and transportation, along with good government, were the first and often the only subject of any speech outlining the basic requisites of economic progress. Now, much elementary mass-production industry over the world does well with subliterate workers because they have the eagerness and discipline that come with escape from the economic privation and social isolation of primitive agriculture. Much labor-intensive industry has therefore moved to the new industrial lands, where muscle and uncomplaining diligence are the prime requirements of economic success.

In the advanced industrial nations, however, education has a central economic role. The modern economy requires a well-prepared, adaptable labor force. The expanding sectors — production based on technology and on the arts and design, the great and growing travel, cultural and entertainment industries and the professions — all must have an educated work force. Education also both prepares and inspires the innovators who respond to the interests and diversions of an educated population. Education makes education economically essential.

It is, to repeat, the economic contribution of education that is most often stressed. Educators make their plea for financial support on economic grounds — on their own special contribution to economic performance. However, a line must here be drawn. The good society cannot accept that education in the modern economy is primarily in the service of economics; it has a larger political and social role, a yet deeper justification in itself.

For one thing, education has a vital bearing on social peace and tranquillity; it is education that provides the hope and the reality of escape from the lower, less-favored social and economic strata to those above.

A measure of social and economic stratification is inevitable in the good society; the complete elimination of a class system is almost certainly impossible. Social decency and political stability require, however, that there be a recognized and effective chance for upward movement, escape from the lower levels to the higher. If this does not exist, there is the certainty of social discontent, even the possibility of violent revolt.

In the United States it was the upward movement of the once-rebellious Irish, Italian and Jewish migrants and minorities that allowed these ethnic groups to exchange troubled, angry, sometimes criminal insubordination for peaceful participation in the society, advancing on to political and economic leadership. For upward escape, either by the individual or by his or her children, education is the decisive agent. The ignorant are held to tedious, repetitive or otherwise burdensome toil and, on frequent occasion, to no work at all. With education and only therewith comes improvement; without it there is none, and the plausible recourse is to crime and violence. A case could be made, and perhaps should be made, that the best in education should be for those in the worst of social situations. They are the most in need of the means for escape.

In the good society there are two further and vital services of education. One is to allow people to govern themselves intelligently, and the other is to allow them to enjoy life itself to the fullest.

Self-government, democracy, no one can doubt, is a demanding thing. An elementary agricultural economy required little from the state; a relatively unsophisticated intelligence was serviceable for both government and governed alike. With economic advance and accompanying social responsibility, the problems facing government increase in both complexity and diversity, perhaps not arithmetically but geometrically. There must then be either a knowledgeable electorate intellectually abreast of these issues and decisions or a more or less total delegation of them to the state and its bureaucracy. Or there must be surrender to the voices of ignorance and error. These, in turn, are destructive of the social and political structure itself.

There is no novelty in this last. All democracies live in fear of the influence of the ignorant. In the United States, from experience with Huey Long, Gerald L. K. Smith, Father Coughlin, George Wallace, the more extreme of the religious fundamentalists and in recent times the militias, it is known that a certain percentage of the population is available to support virtually any form of political and social disaster. It is education and education almost alone that keeps this minority to a manageable number.

But there is a further and less evident relationship between education and democracy. Education not only makes democracy possible; it also makes it essential. Education not only brings into existence a population with an understanding of the public tasks; it also creates their demand to be heard. Illiterate men and women, especially if scattered over the landscape in subordinate relationship to landlords, can rather readily be kept silent and under authoritarian control; this is well recognized. Such is not possible with an educated and thus politically concerned and articulate citizenry. The point is readily established in the modern world. Here, there is no well-educated population that is subject to dictatorship or, at a minimum, is not in a measure of revolt therefrom. Dictatorship of the poor and illiterate, on the other hand, is a commonplace.

Traditionally we think of democracy as a basic human right. So it is. But it is also the natural consequence of education and of economic development. That is because there is no other practical design for governing people who, because of their educational attainments, expect to be heard and cannot be kept in silent subjugation. So, to repeat: education makes democracy possible, and, along with economic development, it makes it necessary, even inevitable. And it has a further reward.

Education is, most of all, for the enlargement and the enjoyment of life. It is education that opens the window for the individual on the pleasures of language, literature, art, music, the diversities and idiosyncrasies of the world scene. The well-educated over the years and centuries have never doubted their superior reward; it is greater educational opportunity that makes general and widespread this reward.

It was anciently assumed, and is still in large measure assumed, that the offspring of the economically and socially privileged would have access to the best education and its enduring advantages. And that they would pay, often handsomely, for it. It was this that brought into existence the private schools and colleges in the United States. The larger life that was so accorded justified the cost.

In modern times, as this rather obtrusive fact has achieved wide and critical recognition, there has been an effort to give an aspect of democracy to these once forthrightly privileged institutions by providing scholarships and financial support to the economically and socially less favored. To the same end, but with a wider service, has been the development in the United States of the public system of higher education — state universities, which, in general competence, are the best of their kind in the world. This too, however, is a greater favor to the more fortunate community. The poor do not have the same access to public institutions of higher learning because inferior, underfinanced elementary and secondary schools, especially those in the larger cities, deny them this opportunity. Here in the United States we have perhaps the most brutal form of social discrimination: some, as a matter of course, are awarded the full enjoyments of life; many are not.

The role of education in the good society is obvious from the above. Every child must have access to and be required to receive a good elementary and secondary education; he or she must also be subject to the discipline essential thereto. Compulsion and discipline are both necessary; the good society does not allow to the very young liberty of choice as between diligence and juvenile distraction. Thereafter as to higher education there must be full opportunity for achievement so far as aspiration seeks and ability allows. For all this, public resources must be available. There is no test of the good society so clear, so decisive, as its willingness to tax — to forgo private income, expenditure and the expensively cultivated superfluities of private consumption — in order to develop and sustain a strong educational system for all its citizens. The economic rewards of so doing are not in doubt. Nor the political gains. But the true reward is in the larger, deeper, better life for everyone that only education provides.

Private and religious schools, colleges and universities are, of course, encouraged; they are an expression of an essential freedom in the free society. They must not, however, be a design for according a better education and superior educational opportunity to those who are able to pay.

The prestige and the income of the teaching profession must reflect the high importance of education in the modern society. Education must both attract and celebrate the best. On two very practical matters all with a concern for the good society should conscientiously reflect. One is the ease and abundance with which money is available for the television that children now so intensively watch as compared with the money provided for their schools and the pay of their teachers. The other is how readily resources are available for the military as opposed to resources for the educational establishment. This is a subject for further emphasis.

10

Regulation

THE BASIC PRINCIPLES

T HE MARKET ECONOMY is based on the un-
planned, uncontrolled response of individual producers and of corporations, small or large, to the
will and the purchasing power of the consumer at home
and abroad. The purchasing power that drives this mecha-
nism originates from the productive activity to which the
purchasing power responds. A closed circle. This is the
essence of the market system. With the collapse of socialism in Eastern Europe and the Soviet Union and its
modification in substantial measure in China, there is no
other. A central question, as all know, is how much this
economic entity, this machine, functions independently
and how much it requires support for, and restraint on, the
purchasing power — the effective demand — that em-
powers the system. Additionally and urgently there is the
question of what guidance and control this machine must
have so that it will serve and not impair the public interest; specifically, what government regulation is needed.

This last ranks as one of the most contentious social and political issues of the time. The conflict is between those who support the autonomous, self-motivated operation of the system and, in particular, the pecuniary interests of those so engaged, and those who see the evident need for intervention to arrest socially damaging or deeply self-destructive tendencies. In recent times in the United States a massive ideological attack has been mounted on public regulation in and of the economy. This again is an escape from thought. There is no specific rule; decision, as on other matters, must be made on the merits of the particular case.

There are four factors that force public intervention and regulation. There is, first, the need for contemporary and long-run protection of the planet, regulatory requirements commonly described as preventing environmental destruction. These are the subject of the next chapter. Second, there is the need to protect the vulnerable among those employed in the productive apparatus from the adverse effects of the economic machine. With this an earlier chapter has dealt. Third, there is the more than occasional propensity of the economy to produce and sell technically deficient or physically damaging goods or services. And, finally, the system incorporates within itself tendencies that are self-destructive of its effective operation. Each of these factors, to repeat, produces a sharp conflict, with ideological overtones, between those who see the system as a fully independent force and themselves as deservedly rewarded thereby and those who advance the case for protective or corrective action.

One further word is needed beyond what has been said on the protection of the worker. In the good society provision for health insurance must go with employment; this has been one of the civilizing steps of modern times. Similarly, care and compensation for work-related injury or damage to health are essential. As is the assurance of a safe workplace. And there is continuing and urgent need, as already stressed, for the income protection of workers in small or small-unit enterprises, specifically the service industries. In general, however, the role of government and public regulation in the field of labor relations has diminished in modern times. Once in the United States membership on the National Labor Relations Board was a source of substantial public attention. Now it is a design for deep anonymity.

At one time the economic machine was the source of the simple, stark essentials of economic life — food, clothing, shelter, fuel, transport, basic materials. Were any of these subject to monopoly control, physical deprivation and suffering followed. From this circumstance came antitrust legislation and other designs to protect the often painfully impoverished consumer against producer-exercised power.

Today, economic change and higher living standards have both diminished and increased the need for product and producer regulation. The greater globalization of economic life, to use the now fashionable term, has lessened the threat of monopoly power and consequent exploitation. Instead of three automobile manufacturers in the United States — oligopoly — there are now numerous competitors, foreign and domestic. Not IBM alone but many sources of computers and their software. So in other countries.

More generally, in an increasingly well-furnished economy there are, given its nature, a great range of choices, each of diminished urgency. Anyone can be allowed his or her own error in deciding between a Cadillac and a Mercedes-Benz. Or between lesser vehicles. Or between differing offers of designer jeans or alternative breakfast foods. In much of the modern consumer movement attention is addressed to the comparative utility of competing products, all of more or less equal merit or with differences of no great effect. The monopoly power of a single producer is no longer relevant.

There is, however, another side. As affluence decreases the need for regulation, so it also increases it. Before automobiles there was no problem of vehicle safety. Nor was there need for highway speed regulation or urban traffic control or action against drunk driving. Or for the control of pollution, as the next chapter will tell. The case is the same with many other products, from toys to asbestos. Modern electronic communications have also introduced a new and contentious area of regulation. And there is a larger and more diversely urgent problem in protecting the buyer from innocently false or overtly dishonest claims. This is especially necessary in the matter of health and medicine. As people become well supplied with the artifacts, physical comforts and diverse enjoyments of life, they turn more and more to what may seem to enhance and extend life and its psychic rewards, and there is an eager industry available in response. Regulation here is a well-recognized necessity. There must be protection against damaging drug and medical use and intervention.

The most urgent and most debated area of regulation has to do with that which affects the operation of the economic machine itself. Adverse conduct here can be deeply damaging, but even when it is visibly destructive, action to correct it can be strongly resisted.

The economic system operates effectively only within firm rules of behavior. The first is common honesty — truth must be conveyed as essential information to investors, the public at large and, as already specified, to consumers. In the field of finance, however, it is especially likely that, misconduct being both remunerative and damaging, this will not occur. Regulation must, accordingly, prevent false or misleading reporting as to business performance and earnings and as to investment prospects. And there are numerous other designs for bilking the minimally informed or mentally innocent. A specific and recognized need is to control insider trading, the use of privileged information. Also open to discussion are large-scale or hostile takeovers when they load the target company with unmanageable debt, and recurrent land and securities speculation with its inevitable and economically depressive aftermath. It must be recognized that from few matters has modern society more suffered than from the excesses and errors of what is now called the financial community, although it once had the more luminous sobriquet of high finance.

The uncertain association of money and intelligence has already been suggested. In the financial world the good society must assume less than perfect performance, especially as each generation returns with enthusiasm to the derelictions and frequent insanities of the one before.

All this, along with environmental protection, is the essential framework of government regulation in the good society. There is, to repeat, no common rule favoring or opposing regulation in the large. Again there must be no escape into ideology from thought; all depends on the specific case within the larger context. It is in the nature of the system that its productive process or its products can have harmful social effects. And there can be longer-run effects that are currently invisible or enthusiastically, even righteously, ignored but that are potentially disastrous. While the need for some regulation, perhaps much, is diminished by modern affluence and the choices it provides, it is also increased by affluence. And there is decisively important regulation that is essential for the effective performance of the economic system itself. As no one can stand in favor of regulation per se, no one can take a general stand against it. If there is a rule, it is only that when a specific regulation is being considered, there should be a search to see if self-serving pecuniary interest is the motivating factor in the argument.

11

The Environment

THE GOOD SOCIETY has three closely related economic requirements, each of which is of independent force. There is the need to supply the requisite consumer goods and services. There is the need to ensure that this production and its use and consumption do not have an adverse effect on the current well-being of the public at large. And there is the need to ensure that they do not adversely affect the lives and well-being of generations yet to come. The last two of these three requirements are in frequent conflict with the first, a conflict that is strongly manifest in everyday economics and politics. The common reference is to the effect on the environment. Here, in brief, are the issues involved as they are defined by the good society.

The production of goods and services is a problem that in the fortunate countries of the planet has been extensively solved. There is still the question of the stability of its performance and of how its revenues and rewards are distributed, questions previously addressed. But the ability of the modern economy to produce in profuse supply what the consumer wants and needs is without question. As told, it does this far beyond any independently established demand. A vast and energetic advertising industry and the persuasive power of modern communications, especially television and radio, are now necessary to instruct the individual as to his or her desires and thus to promote the resulting consumption.

The environmental problems emerge from the impact of this production and consumption on the contemporary health, comfort and well-being of the larger community. And they come from their future effects, including the exhaustion of the natural resources that are now so abundantly available and consumed.

The manifestations of contemporary damage are distressingly familiar — air and water pollution, the large and growing problem of waste disposal, the immediate danger to health from the products and services dispensed and the visual pollution from the intrusion of production and sales activity, particularly retail sales activity, on the urban and rural landscape. Not infrequently, bad health and visual pollution go together. In their great steel-producing days, Pittsburgh, the English Midlands and the Ruhr were both dangerous as to health and hideous as to aspect.

The long-term as distinct from the contemporary effects are many: the delayed damage of air pollution, the most discussed examples being global warming and the greater incidence of lung cancer and emphysema; other disastrous climatic changes, as from rain forest depletion; the exhaustion of mineral, petroleum and other resources on which current consumption depends; and, more distantly, as the population grows and urbanization continues unrestrained, the exhaustion of relevant living space itself. There are also exceptionally complex issues that have to do with the protection of wild life and, especially in the United States, with the protection of public lands and parks from aggressive commercial invasion or expropriation.

Such are the contemporary and longer-run environmental effects of our consumer economy. How does the good society react?

The first requirement is strong and enlightened citizen concern. Environmental protection produces no immediate economic return; for it to gain support and achieve its goals, there must be alert and persistent public and political expression and action. Here, the present situation is not entirely discouraging; environmental issues currently inspire a widespread and often effective public interest. This is vitally important; in the good society it must be strongly encouraged.

The economic and political situation must also be clearly understood. As earlier indicated, environmental concerns, both those which are contemporary and those affecting future generations, especially the latter, are inherently in conflict with the motivating force of the market economy, which is immediate, foreseeable return to the producing firm. This, in turn, commands the energy and intelligence that empower the economic system both physically and mentally. Any intrusion on this system and its motivation is seen as socially and economically damaging, and especially so by those who experience it. A sacrifice of freedom of decision and profit in order to protect the larger community or its unborn children is held to be an abridgment of the very freedom that produces economic success. The conflict is not lessened by the fact that government — the state — is the principal instrument for protecting both the present and the longer-run environmental interests. By an attack on the government as an ill-intentioned intervening force, environmental legislation and needs can be successfully thwarted, a strong tendency as this is written.

The conflict between the contemporary and the eventual public effects of the consumer economy and the short-run dynamic of the economic system is a matter of everyday observation and debate. The electric utility brings needed power and light to its users. In so doing, it contributes to atmospheric pollution, to problems of fuel waste disposal, quite possibly to eventual resource depletion and, in specific cases, to the threat of nuclear disaster. The automobile around which, to a remarkable extent, the modern consumer economy revolves contributes similarly to air pollution and, in the occupation and use of street space, to urban environmental degradation. And there is again the long-run effect of its fuel consumption on petroleum resource depletion. For a long time the United States supplied itself with gasoline from its own oilfields. Now, after only a few decades of more and more automobiles and trucks and their increased use, it is dependent on, possibly held hostage to, the Middle East. On some admittedly distant day the resources there will also be severely limited or gone. The modern economy has a large construction industry; this can mean the progressive destruction of the forests, the endangering of the animal, bird and other species there residing and lessened recreational and visual enjoyment of the wooded areas.

The effect of a vigorous economy on visual pollution calls for a special comment. Mention has already been made of the dark satanic mills in the great steel-producing centers. Important in the modern world are the visual abasements of roadside commerce by the advertising that sustains consumer demand. In this area the long-run effect has already arrived. The American countryside is far less beautiful for the casual traveler than it was a hundred or even fifty years ago. And the case is the same, if in lesser measure, in the other rich countries. Britain, France and Switzerland, to their credit, are more protective of their landscape. Over much of Japan, which was once beautiful, the roadsides are now modeled closely on the scenic grace of Jersey City, New Jersey.

The good society does not deny the existence of the conflict between basic economic motivation and contemporary and long-run environmental effects. It seeks to resolve the conflict in a rational way, but resolution will not come from either prayer or public rhetoric. There is no escape from the role of government; it is for the larger community interest and its future protection that government and governmental regulation exist.

The market system and its incentives are an accepted part of the good society; this is not in doubt. But there is no divine right of free enterprise, of free choice, for the producing firm. Or for its consumers. The larger community interest must be protected, as also the future climate and well-being, and there must be concern as to depletable resources. Since automobiles must be built, have fuel and be driven, and other consumer goods and services must similarly be supplied and utilized, a compromise between the current financial and the longer-range public interests is essential and inevitable. As a broad rule, however, this compromise must favor the larger community interests and the interests of those to come. That is because the business and political voice and money are allied with the current economic power — with the firms that produce the goods and services, their lobbies and their captive or susceptible politicians. The community and the longer public future draw on less specific support.

In the good society the environmental concern must have a strongly motivated constituency endowed by its members with the necessary financial resources. There must also be a presumption in its favor in public discussion and political action. Economic reward — profit — and the religious fervor that regularly defends its unfettered acquisition lie with the opposition, with the producers, the suppliers and their marketing and advertising specialists. Public balance requires that there be those who champion effectively and cogently the contemporary and long-run environmental case. They should not be immune from intelligent criticism, but the weight of public opinion and political support must always be on their side.

12

Migration

NEXT ONLY to what benefits the poor, perhaps the most controversial problem with which any modern society must contend is that of migration — in practical terms, the movement of people from the less to the more advantaged countries and their effect on the social and economic life of the lands to which they go. So it is in the United States, Canada and Western Europe, and so in the future it could quite possibly be in Japan.

As the modern economy develops, it comes increasingly to rely on immigrant labor from abroad. Without this labor supply, there would be grave economic disorientation, even disaster. Yet there is a strong current of thought, or what is so described, that deeply deplores immigration, is deeply resentful of the migrants and campaigns ardently against their entry and continuing presence. There are today no industrially advanced countries, Japan being the exception, on which the question of migration does not obtrude as a major political issue.

*

The basic circumstance is not in doubt. In Western Europe, which is the clearest case, a wide range of industrial and service enterprises depend on immigrant labor. In Germany automobiles would not be assembled, other industries staffed, a great variety of services rendered, were it not for Turkish workers, those from the former Yugoslav states and from elsewhere in Eastern Europe. There would be similar difficulty in France were it deprived of the North Africans. Italian industry once depended on workers from its own south, the relatively impoverished Mezzogiorno; now it too reaches out to Africa. Spain, which for long supplied workers to other European lands, now relies in some measure on Africa as well. Britain has replenished its industrial work force and staffed its service industries, including numerous small retail establishments, with former residents of its erstwhile empire.

In the United States successive waves of immigrants, first from Europe, later from Asia and Latin America, have gone into both industry and agriculture; there would now be little fruit, few vegetables and fewer canned goods in our stores at affordable prices in the absence of migrant workers.

The controlling fact, which has been rarely remarked in economic literature, must be made clear at the outset: there is a problem with the word "work." It is used to characterize two radically different, indeed sharply contrasting, commitments of human time. Work can be something that one greatly enjoys, that accords a sense of fulfillment and accomplishment and without which there would be a feeling of displacement, social rejection, depression or, at best, boredom. It is such work that defines social position — that of the corporate executive, financier, artist, poet, scholar, television commentator, even journalist. But work also consigns men and women to the anonymity of the toiling masses. Here it consists of repetitive, tiring, muscular effort replete with tedium. It has often been held that the good workman enjoys his work; this is said most frequently, most thoughtfully, by those with no experience of hard, physical, economically enforced toil. The word "work" denotes sharply contrasting situations; it is doubtful whether any other term in any language is quite so at odds with itself in what it describes.

Additionally there is the vitally important matter of compensation. As a broad rule, those who most enjoy what they do, who find work most agreeable, also get the highest financial reward; those for whom work is the most repetitive and physically exhausting get the lowest or some approximation thereto.

A basic feature of the good society, one that has already been noticed, is the opportunity it affords for upward economic and social movement. The major incentive is the movement from, as it may be called, real work to what is only denoted as work. This creates a vacuum at the bottom, which it is an essential service of migration to fill. There is need for this constant refreshment of the labor force in the area of monotonous, nonprestigious toil, and this need is met by people in escape from the yet more tedious and yet more ill-paid employment of the poor countries or by those with no employment at all. For them the lower pay and strenuous work that are available in the affluent lands are still far better than anything they can find at home.

Migration that is so induced is not exclusively an international matter. The reliance of northern Italian industry on willing labor from the south of Italy has already been mentioned. An even more dramatic movement was that of the poor in the Appalachian Mountains of the eastern United States and of the former plantation sharecroppers and rural workers in the South to northern industry and service occupations. This migration was also not without reaction; the movement of blacks and poor whites into the large northern cities created social tension and occasioned adverse comment.

The reaction to foreign immigrants, as also to internal migration, comes partly from the belief or, in any case, the assertion that the newcomers are taking jobs that properly belong to established workers already in residence. That many of the migrants, if not most, take employment for which the resident workers are not available or that they no longer seek goes unmentioned. A further, much cultivated negative reaction is ethnic and social — the newly arrived are thought to bring a different and presumptively defective racial, religious, familial, hygienic or civic culture to the established community.

Among the industrially advanced countries the United

States has been the most socially tolerant of immigrants in the past. A large and persuasive literature praises the contribution of the migrations that have occurred to the American society, the benign and affirmative nature of the melting pot. But that view applies extensively to earlier migrants. Toward current arrivals there is a strongly negative attitude that is manifested in political oratory, discriminatory legislation and occasional bursts of community hostility. And the case is the same in varying degrees in the other economically advanced lands.

The extreme, and in some ways the special, case of resistance to migration is Japan. There, insular geography and a strong sense of cultural identity have combined to limit the flow of immigrants. In the future this could restrain Japanese industrial development by denying the country an urgently needed working force at the real-work level. There is already some indication of this in recurrent labor shortages and a certain amount of highly informal immigration that is not accorded any legal recognition.

It would, of course, be a serious error to confine consideration of the problem of migration to the working masses — those in pursuit of real work. In a world community in which there are close links between business and finance, art, literature, entertainment, intellectual and scientific pursuits, there is a large and growing exchange of business, academic and cultural talent, or what is so described.

Here again the difference in public attitude as between work and real work: the migration of the socially, culturally and economically well endowed encounters no serious objection. On the contrary, it is greatly praised and, in practical fact, is subject to few legal constraints. As in the noted Cockney verse, "It's the poor what gets the blame."

How, given the depth and diversity of the problem, should the good society respond?

The good society must accept that two worthy and commendable objectives are here in conflict. (A not unrelated situation will presently be seen in the national welfare focus of the modern state and the increasing internationalization of economic, social and cultural life.) For the poor of the world, migration is the most evident escape from privation and suffering. And concern for fellow beings wherever located is or should be on the conscience of all. Accordingly, the poor should be granted the opportunities and enjoyments of the more favored lands.

At the same time governments have an undoubted duty to their own people — employment, welfare support, health care, much else. The larger world obligation must be reconciled with the local, that is to say, national, responsibility.

Any resolution of the conflict between the two must accept and explicitly favor a steady flow of migrants. This works to their advantage and equally to the advantage of the receiving country. Since this is already admitted and celebrated when the émigrés are at the higher levels of talent, it is important, even urgent, and a mark of civilized behavior, that the service rendered by those who do real work be similarly recognized and applauded.

The last point must be emphasized. The tendency to

see the poorer immigrant as an intruder and in some measure as a burden is something the good society rejects. It sees the immigrant worker in the full light of the service he or she performs. It is understood and accepted that life in the advanced countries would be difficult without a steady foreign contribution to what, admittedly, are the lower, more arduous levels of the labor force. Accordingly, those coming and so serving should be both welcomed and encouraged and, needless to say, should encounter no discrimination or hostility based on race, color, language or cultural difference.

There must also be an opportunity for the upward economic and social movement already emphasized, and especially so in the succeeding generations. One returns to the main point: a liberal immigration policy in the good society serves those who seek to come, and it serves no less substantively those who are already there.

An important question remains, however. Given the responsibility of the national state for its own working force, should migration be at least controlled in its favor?

The practical answer is yes. There need be no effective limitation on international or internal movement in the higher brackets of achievement — on the immigration of literary, artistic, scientific, technological, athletic and like talent, those engaged in business and, quite possibly, those primarily committed to leisure and its enjoyments. And there must be an open door for those in escape from overt political aggression, as, in principle, is currently the case. There must also, to repeat, be no discrimination, actual or implicit, as to ethnic identity or race.

But as to those headed for the real work, the unfavored toil, of the more fortunate countries, admission must, no doubt, be related to the availability of jobs. No country can be burdened with a large surplus of immigrant workers beyond the demands of the lower levels of employment. Some countries, Switzerland being the leading example, have arranged their immigration policy to prevent this, and with success. The case of the United States is far more difficult, and such a calculation may be beyond both the statistical competence of the government and its control of its borders, especially that with Mexico. There, with slight effort and some persistence, workers can manage entry without restraint by the immigration authorities. As in other matters contributing to the good society, and perhaps more than most, immigration policy cannot achieve perfection. Nor would that standard be approached, as has been suggested in much modern discussion, by denying benefits, including education, to those already arrived.

The larger answer is for the good society to recognize the beneficent role of migration in general and to act and react accordingly. The national community is enriched by those of foreign culture and sophistication and by the exchange of ideas and talents that a liberal immigration policy allows. And there is specific economic advantage to the rich lands from the movement of workers from the poor countries for the real work that in the affluent world all but the avowedly eccentric seek to escape.

13

The Autonomous Military Power

IN THE MODERN POLITY the public living standard, as well as the socially necessary restraints on private action, is the result of democratic process and decision. This arrangement is far from perfect or peaceful; every day the press and the media in general report the political pressures and conflicts that surround these matters and how they are resolved. And there is a basic, overriding problem evident in this discussion. The private living standard, as previously stressed, is the beneficiary of enthusiastic, often relentless advocacy; that is the function of all salesmanship, all advertising, all product and service promotion. By contrast, the public living standard — schools, parks, libraries, law enforcement, public transport, much else — has no such support. The consequence, one that is wholly familiar, is expensive television and meager schools, clean houses and dirty streets — what I once characterized as private affluence and public squalor. But within the allocation to public purpose itself there is an especially egregious error in resource distribution.

That is as between military and civilian needs, and it is the result of a serious failure in the democratic process.

In the United States the decision as to public expenditure is made through a combination of legislative and executive power. The defining and controlling factor in all public action is the money thus provided. Much may be specified by law; if the money is not available, not much will happen. Both the legislative and the executive branches of the government are subject, in turn, to election by the citizenry at large, the determining force of money thus becoming a direct and effective manifestation of democratic authority. Or so it is assumed.

In fact, there is one major exception to this exercise of democratic control, and that is the military power. This has often been true in other countries, and especially in those, as they are called, of the Third World. The United States, however, is presently the particularly clear example. The American military establishment effectively and independently decides on its own budget, on the extent and the use of the money it receives.

The claim on public funds by the military and its plenary power over their disposal are routinely accepted in the executive branch of the government. It is tacitly agreed that civilians in nominal authority do not tangle seriously with the military. This is a force with which budget directors, presidential appointees, even the President himself, do not effectively contend. What has just been said is so thoroughly acknowledged that it is little discussed.

The power of the military — its ability to assert need and acquire the requisite financial support — is evident in even greater measure in the legislative branch. There, subject to minor symbolic changes, the military budget is voted automatically and, as this is written, with an annual increase. It is not even thought necessary that some military needs — intelligence operations, sophisticated weapons — be made known to the legislative body as a whole.

Effective also in the Congress is the financial and political power of those industries which produce the weapons; legislators do not vote readily against the employment the weapons makers provide to their constituencies. More important, perhaps, is the reassuring manifestation of patriotism in supporting the armed forces. The authority of the military establishment has thus become complete, a circumstance that is generally conceded.

If there is full, unquestioned power to obtain the requisite funds and equally unchallenged power to decide as to their use, that power is then total. Democratic control has been effectively set aside. This, with no exaggeration, is the present situation as regards the military in the United States. The resources it commands are not determined by need; that is not seriously asserted in knowledgeable circles. Distinguished former Army and Navy officers, notably those associated with the Center for Defense Information, regularly question the necessity for particular weapons and force levels. The end of the Cold War was an impressive fact; it did not affect the continuing claim of the military establishment on money and the executive and legislative support that provides it.

Over the centuries in many countries there has been strong assertion of the military authority. Weapons and the disciplined ranks possessed of them have stood as a threat to civilian control. With this, those who framed the American Constitution were much concerned; thus the designation of the civilian President as the Commander-in-Chief, the ultimate authority over the armed forces. From the modern-day power of the military over its own financial resources and the use to which they are put has come, in substantial measure, what the Founding Fathers most feared.

The situation in the United States is not, as noted, unique. In Central and South America, in much of Africa and Asia, the armed forces have been, and in many countries remain, independent of the civilian control embodied in democratic government. Or they have replaced it. The United States, trapped by military expenditure unrelated to military need, has developed an unfortunate Third World aspect. It has surrendered substantially to what Dwight D. Eisenhower, General of the Army and Republican President, was moved to warn against: the emergence of an independently powerful military-industrial complex.

The good society does not concede authority to the military power. This is not because of the danger, much feared in the less fortunate lands, that it will replace the civilian government. Political structures in the United States and other industrially advanced countries are too well entrenched for that. Rather, it is because the modern military power is not beholden to the larger public interest, urgently and solemnly as this is avowed; it is governed by its own interest, which, moreover, can be intensely damaging to the larger public needs and goals.

The adverse effect of the American military power on policy was especially evident in the early 1990s. Then, the collapse of Communism and the disintegration of the Soviet Union brought an enormous need for international resources, dollars in particular, to finance and ease the transition of the former Communist regimes to something approaching a market economy. Such help had been made available through the Marshall Plan in the generally less difficult move to a peacetime economy in Europe after World War II. Much suffering and much potential for political disorder were thus mitigated or avoided. So it would have been in the former Soviet Union and its onetime acolytes in Eastern Europe. Instead, the military power in the United States remained committed to using public resources to guard against a military threat that now, admittedly, had disappeared. The needed help to the erstwhile Communist countries was not forthcoming in any adequate volume.

Similarly on the domestic scene, there were and there remain the insistent claims of the poor and the impoverished, especially in the larger cities. Whatever the earlier foreign threat, there was and there is still the danger of an abrupt and violent challenge to domestic tranquillity. The military power continues to control for its own purposes the resources that, if used for basic income support, job creation, housing and drug counseling, would ease the crisis in the inner cities.

We have here perhaps the largest and most evident intrusion on the standards of the good society. As has already been emphasized, there must be a constant watch over the claims, particularly those of the influential and the powerful, on public expenditures. No one should doubt that in public services and outlays there is a well-established bias in favor of the fortunate. But the military and its needs must be recognized as the special case. For all who seek the good society the primary concern must be that the autonomous military power that now exists be brought under effective democratic control. To this end, the strongest political voice and action must be directed.

The intrusion of the military on the modern polity and economy, its claim on public resources, is not that of individuals. From time to time unusual men do appear and achieve national prominence — chairmen of the Joint Chiefs of Staff or commanders on some faraway battlefield. The power of the military comes, however, from the power of organization, mass organization. This can have a character and purpose and a claim on resources that far transcend the authority of any one person. In common terminology, this is bureaucracy, of which the modern military establishment is, in many respects, the extreme case. The role of bureaucracy and bureaucratic power in the good society and its frequent conflict therewith is my next concern.

14

The Bureaucratic Syndrome

ORGANIZATION, with its power and, all too often, its weakness, is a central feature of modern life; the bringing together of individuals in a hierarchical structure of command and cooperation in pursuit of a common purpose is indispensable to the effective operation of every aspect of contemporary existence. In the government there is the large public agency; in the private sector, the great corporation.

Both public and private organizations are subject to a measure of critical scrutiny and comment. The public agency is regularly condemned as a "bureaucracy," the word having a markedly negative overtone; faithful, intelligent and essential civil servants are often denigrated as "bureaucrats." The terms are also used, if not as aggressively, to characterize the less-than-effective administrative apparatus of the large corporate enterprise. The good society must recognize and contend with what has become a bureaucratic syndrome in both the public and private sectors. Its basic flaws are two, each of which has a life of shadowed recognition.

*

The first and most evidently adverse tendency of organization, large organization in particular, is that discipline is substituted for thought. Discipline is inescapable; there must be acceptance and willing pursuit of a common goal, for it is this that makes organization effective, even possible. The individual who conforms fully is commended in highly relevant metaphor as "a good soldier." At the same time there is no doubt that creative thought is suppressed and often replaced by the disciplinary process. The man or woman of independent view — who identifies weakness or error and sees or foresees the need for change — may well be considered uncooperative, irresponsible, eccentric. In a favored government expression, he or she "is not useful." In all organization there is always this basic conflict: on the one hand, the very practical need for cooperative acceptance of the established procedures and purposes; on the other, the need to question those procedures and purposes as error or events call for change. And also the ability and the will to urge and effect such change.

This conflict, as just observed, is common to both the public and the private sectors. In the case of the government department or other public agency, the static tendency of organization is well recognized. The everyday reference to bureaucracy and to bureaucrats generally reflects a negative attitude toward the public service being rendered, including its politically disagreeable effect or cost for those so speaking. But it can also describe obsolete, irrelevant or incompetent action. It is the task of the good society to distinguish between the two.

This is especially the case as regards foreign policy, a matter to be discussed in the next chapter. In the United States the Department of State and to some degree also the Central Intelligence Agency and the Pentagon administer not legislation, not services, not public programs, but policy. That, in turn, invites and requires belief. Belief is essential; without it, the conduct of the policy would be impossible. But although controlling circumstances change, belief, once established, does not. The militarization of American foreign policy in the 1950s led in later decades to such disastrous misadventures as the U-2 over the Soviet Union, the Bay of Pigs, the paranoia about Communism in Central America (and Cuba) and the Vietnam war, all testaments to the inherent rigidity of administered belief.

In the modern great business enterprise the bureaucratic syndrome also has its readily identifiable presence. A comfortable and disciplined culture resting often on past success takes the place of innovation and change. In the United States the steel, automobile, computer, airline and retailing industries have all provided formidable examples of this tendency in recent years. Other countries have similar cases. Unlike public organization, however, the private enterprise can be subject to the shock effect of eventual financial trauma. It can fail and go out of business, be taken over by another, more successful firm or be forced by external financial threat into self-reform. Bureaucratic stasis is, however, an omnipresent fact in the great private, as in the great public, organization.

*

The second and related feature of all organization was earlier suggested — an internal dynamic that leads to the relentless proliferation of managerial and other personnel. The controlling circumstances that govern personnel policy in both sectors of the modern economy are simple and wholly obvious, but they normally go uncelebrated, with the tacit consent of those involved. There is, first, the desire of anyone in a position of hierarchical responsibility to want a seemingly sufficient body of supporting staff. The workers so acquired have, in turn, their own desire and apparent need for assistance. Specialization then adds to the need; there must be personnel of suitably varied knowledge and competence. The whole process, as indicated, has a dynamic of its own.

And there is more. From numerous and suitably deferential subordinates come both the reality and the enjoyment of power. Also prestige within the organization and a claim on higher pecuniary compensation. An accepted measure of an individual's worth is the number of people over whom he or she presides: "How many does he have under him?" To add subordinates is thus to enhance in the most visible way position, prestige and pay.

There are, of course, efforts to limit the expansive process. To this end budgets are prepared and budget limits imposed. These, however, can be largely symbolic. In all great organizations a strong and even irresistible tendency is to add managerial, technical, professional and other employees. Only as one gets to the shop floor in the industrial corporation — to, as significantly they are called, the working levels of the enterprise — is the proliferation dynamic held in check. Only at these levels — the worker on the assembly line, the elementary clerical staff — is there a close, continuing assessment of needed workers to product.

It is the wonderfully perverse tendency of economic behavior that results are frequently more visible than cause. That is the situation here. In the private sector the inner-generated growth of personnel just described develops, over time, a quiet life of its own. Then, after recognition of the process, comes the result: a recurrent and often much publicized program to delete nonfunctional workers. These efforts — usually called corporate downsizing, never forthright firing — are regularly reported on the financial pages. Not asked is what those so discarded were doing before, why they were originally added to the organizational structure, how the organization will function without them. Also unmentioned or little mentioned is the personal disaster of those so released, many into enduring unemployment, all into a mental condition of proclaimed uselessness.

That public organizations — the Pentagon, the State Department, the Commerce Department and others — have the same tendency to nonfunctional proliferation must also be accepted. Organization is organization wherever it exists and under whatever auspices. The question is how the good society should react.

There is no easily defined course of action. The answer lies in accepting that bureaucratic stasis and unnecessary personnel proliferation are the basic flaws of all organization. Of these fundamental tendencies the national charities, the college and university administrations, trade unions, other big organizations, must also be deeply and constantly aware.

The private enterprise does have the virtue, as noted, that competitive incompetence — what perhaps may be called the General Motors–Ford solution — may compel reform. Financial loss, the threat of bankruptcy, has its salutary effect. There is obvious advantage, however, in creative foresight, acting in advance of market-forced disaster. There is undoubted social decency in intelligent action that obviates the need for recurrent and painful downsizing of unneeded staff.

The case of the public agency is more difficult. Here, neither the institutional commitment to established policy nor the proliferation of personnel has the inbuilt remedies that exist in the private organization. Instead, reliance must be on effective and informed management and a willingness to effect change when that is needed. Normally, as has been earlier observed, any attack on bureaucrats and bureaucracy in the public sector is a cover for opposition to the particular service being rendered, the particular law being enforced or the cost thereof. But the problems inherent in the bureaucratic syndrome do exist in the public agency, as in any large organization. Their solution is in the hands of vigilant leadership in the executive and legislative branches, and solution is essential if the good society is to work effectively.

15

Foreign Policy

THE ECONOMIC
AND SOCIAL DIMENSION

T HE ULTIMATE OBJECTIVE of the good society is in the field of foreign policy. There, it seeks lasting peace between nations. Nothing is so important, for nothing so contributes to sorrow, deprivation and death as military conflict. This consigns the young, and now in the nuclear age the civilian older and old, to their sudden and certain destruction and to the inner despair, however concealed, of living with its prospect. In the realm of human intelligence few matters have been so strenuously attempted over time as the justification of war and of warlike achievement. And none has so clearly denied the claim to human progress and civilized enlightenment in the century now ending as the two great wars in Europe, the one great war in the Pacific and the lesser but equally cruel conflicts in Korea and Vietnam. These have been the searing exceptions to the movement toward a good or better society. The prevention of such mass tragedy is, along with a solution to the problems of the poor of the planet, the most urgent task of any society that embraces and seeks to protect all mankind.

Success in the economic dimension of the good society is central to peaceful relations both within and between nations. It is essential for peace and tranquillity within the nation-state; it bears heavily on the relations between states, and increasingly so, as will presently be urged. But this is not all. There remains the need to counter and negate the presumption by the military power that war is an inevitable aspect of human existence, with the consequent demand for weapons of ever more fearsome effect. From this, more than incidentally, also comes the major scandal of our time — the weapons trade to the poor lands, a trade that regularly and abundantly supports military establishments in those countries where there is still an overwhelming need for life-giving food, medical care and education.

Of central importance, however, is the conflict between the social and economic responsibilities of the nation-state in the well-endowed society and the increasingly internationalized economy and polity.

In public affairs the easiest choice is between what is obviously right and what is palpably wrong. And there can also be agreement when both courses of action are perverse as to effect; then there will be a measure of acceptance of the common cost, perhaps suffering. The difficulty occurs when both courses of action seem benign. Each will then have its passionate advocates; there will be a strong element of righteousness in the advocacy. Little so engages mind and speech as believing oneself to be right when others also believe themselves to be right. This is the case when one is discussing the role and obligations of the nation-state as opposed to the rewards from the closer global association of nations. Both are seen as being for the common good, and because of this, conflict between them is inevitable.

Over the last century social initiatives and the larger thrust of history have combined to increase greatly the social and economic responsibilities of the nation-state in the economically advanced lands. Of these readers have been made abundantly aware: social security for the old; unemployment compensation for working men and women; national health care, still an avidly debated matter in the United States but elsewhere generally accepted; the regulation of working conditions, especially those of women and children; a minimum wage; support to education and research; price protection for farmers because of the uniquely rigorous character of free-market competition in their industry.

Additionally, the modern government, as sufficiently noted, can no longer stand by when, in the long-established tendency of the market system, economic performance proves to be imperfect or painfully inadequate. Inflation, recession and unemployment are not now suffered as inevitable; governments are held responsible, and no less if the remedy is partly beyond their reach. In no country, as has been suggested, do holders of public office wish to present themselves to an electorate when economic times are bad. No other consideration, come an election, is thought so important, so decisive.

All of this is to be welcomed. Capitalism in its original eighteenth- and nineteenth-century design was a cruel system, which would not have survived the social tension and the revolutionary attitudes it inspired had there not been a softening, ameliorating response from the state. In recent times there has everywhere been strident oratory, from those in personally comfortable economic positions or addressed to those so favored, that has regretted and condemned the modern welfare state; those so speaking would not now be enjoying a pleasant life in its absence. But there has also been the companion international development, which, while equally benign in effect, is in seeming opposition to the internal social purposes of the nation-state.

The companion development has been the closer association of the peoples and institutions of the economically advanced countries. This includes international trade, which is much discussed. And international finance — the flow of investment (and speculative) funds from one country to another — also much discussed. And the modern transnational corporation, which moves effortlessly across national frontiers. And much more. Travel; scientific research; literary, theatrical and artistic achievements; entertainment, educational pursuits; all increasingly and ever more casually cross international boundaries. All of this the historically alert person must applaud. In the last half of the century that is now coming to an end, it has been in rewarding contrast to the aforementioned wars that so nurtured death and human misery in the first half. It can now be assumed that the favored countries of the planet will live peacefully together because of the internationalization of their economic, social and cultural lives. No one who experienced the earlier world can for a moment be sorry that this is so.

Here, however, we have another modern-day dialectic. The economically guided, welfare-oriented state is good — inevitable. So is the internationalization of economic, cultural and other life. But between the two there is or, in any case, there seems to be inescapable conflict. The internationalization of economic life will, it is feared, threaten the welfare system of the nation-state. Also its cultural and social identity, an expression of national personality and a focus for patriotism to which there can be obligatory, near theological commitment. These attitudes are strongly in defense of the nation-state.

The economic threat of globalism, as it has come to be called, can seem especially urgent. Those countries with better social and working conditions invite competition from lands with lower wages, less effective protection of the economically vulnerable and hence lower production costs. To them the transnational corporation can readily move its operations.

It is possible that this threat is exaggerated: even countries with strong wage and welfare systems — including

Japan, which has something approaching lifetime security of employment — are successful in international competition. Nonetheless, national well-being and particularly the welfare state are seen as threatened by international trading patterns and the transnational business enterprise. This is currently a matter of intense concern and discussion around the world.

There is also the limitation that the international trading and financial system places on movement toward economic stabilization. When there is recession or depression, it traditionally falls to the nation-state to make the necessary stimulative response. Direct fiscal action — tax reduction, public job creation, other fiscal measures to expand demand — is in order. So is monetary action — lower interest rates — for its favorable effect, real or imagined. But the efficiency of these responses is diminished in an international system. Some, perhaps much, of the stimulative effect will be lost in international trade — in increased purchases from, and employment in, other countries. In technical terms, some of the multiplier effect of increased public or private expenditure is lost to other lands. Some of this loss could, in theory at least, be prevented by currency depreciation; exports would then be cheaper, imports more expensive. But this runs counter to a basic tenet of the larger trading community, which is the need for stable currency relationships, extending on, as now in Europe, to the promise of a common currency.

*

The issues here discussed have entered the politics of all the industrially advanced countries; given their pervasive character and urgency, this is hardly surprising. Especially interesting is the schismatic character of the political reaction. On the right, traditional conservatives are caught between a strong commitment to national identity and traditional patriotism and a simultaneous preference for free trade and support for the transnational character of the modern business enterprise. Patriotism, all recognize, is a prime conservative virtue. But present in all reputable conservative thought is the ancient argument for international trade, the maximizing of efficiency in the production of goods and services. While perhaps not an especially urgent matter in the modern economic world, with its proliferating supply of such goods and services, productive efficiency has a powerful hold on traditional belief. National identity is strongly and proudly avowed by the conservatives, as are the economic rewards of internationalism.

On the left there is similar division. Many who support the welfare functions of the nation-state decry the possible loss of employment to lower-wage, lower-cost production in the economically and politically associated countries. At the same time there is applause for the peaceful nature of the larger international community and for closer political, cultural, scientific and economic ties. And there is also on the left the classical commitment to free or relatively free foreign trade.

In recent times there have been dramatic examples of

this conflict. The Maastricht Treaty convention, another step toward closer European association, brought ardent public debate in which traditional political groupings were internally divided. Some on the left were for, some against. Those on the right reacted similarly, a notable case being the British Tories.

An equally dramatic demonstration in the United States was the debate over NAFTA, the North American Free Trade Agreement, and GATT, the General Agreement on Tariffs and Trade. Here too within both the left and the right there were differing and angry views. In the case of GATT there was fear, more eloquently stated than profound, that sovereignty would be surrendered on economic and social issues to the supervising international authority — the World Trade Organization. In Canada a large and articulate minority saw in NAFTA, as also in earlier negotiation, a major threat to that country's longstanding effort to maintain an economic, cultural and political identity separate from that of her large, socially and economically aggressive neighbor.

Eventually both agreements were approved — in the United States by an unnatural coalition of internationally minded liberals and trade-conscious conservatives. Few will doubt the force of the basic conflict involved here. How is it resolved in the good society?

The solution is not difficult; it has the advantage of inevitability. The move to a closer association between the peoples and institutions of the advanced countries cannot be resisted. It is on the great current of history; the social forces involved are beyond the influence of national legislatures, parliaments and politicians. The oratory may oppose it; the tide still will run.

Nor should one wish otherwise. The nation-state's jealous regard for territory, its protection of its own economic interest, the economic power of its national arms manufacturers, the passionate attention to the preservation of its language and cultural identity, were the source of the greatest tragedies of modern times. But in spite of the gains implicit in internationalism, there remains the considerable matter of coming to terms with the inevitable.

Among the advanced countries there must now be effective international coordination of social and economic policies. This begins with fiscal and monetary action, which is essential if the normal sequence of boom and bust, speculation giving way to unemployment, is to be countered. No single country can act effectively and alone. Going beyond this, there must, as earlier indicated, be coordination of national social policies, agricultural policies, measures to meet environmental needs and the other substantive programs of the modern welfare state. The good society must be committed to this coordination, for it is not only the best but the only answer.

The need for such action is recognized in Europe, if imperfectly, by the members of the European Economic Community in Brussels, the Strasbourg parliament, of late rather explicitly by the Maastricht Treaty and most recently in the rather remarkable step taken by the major states to coordinate military procurement. The need for coordination is less recognized in the United States, Canada, Japan and the trading community of the Pacific, and the recognition that there does exist is considerably greater than the practical response.

Presidents and prime ministers now meet at regular intervals primarily to discuss economic issues. Much attention is concentrated on trade relations. Henceforth those assembling must turn to the associated matters — welfare, fiscal and monetary policies and their coordination. These can no longer be left to the different and thus at times self-defeating decisions of the individual nation-states. Trade will, to a great extent, take care of itself; in any case, the returns from negotiation are often invisible. But the effects of the differing welfare, fiscal and employment policies can be strongly felt. It is on these that there must be continuing deliberation and action.

However, this is only the first step; the closer and welcome association that now brings peoples and institutions into international concert must also, over time, bring into being the necessary international organization. The economic and social responsibilities of the nation-state are a transitional phase. The ultimate goal is a transnational authority with the subsidiary powers, not excluding the raising and spending of revenue, that go with it.

In the real world now there are the World Bank and the International Monetary Fund, which in 1995 celebrated the fiftieth anniversary of their conception. Both represent a delegation of economic power to a higher international body. The Bank assumes responsibility for the flow and guidance of investment funds beyond the authority of the individual state, although the commanding role of the United States in Bank affairs cannot be ignored. The Fund, which was initially designed to seek stability in international currency relationships, has gone on to specify domestic budget and expenditure policies that bear upon such stability. Both Bank and Fund have, however, devoted their efforts and resources to the poorer countries; they have not, in any effective way, sought coordination of the fiscal, welfare and other policies of the economically advanced lands. The Fund, additionally, has too often seen the welfare functions of the nation-state not as something to be protected but as something to be sacrificed for currency stability. The Bank and the Fund, nonetheless, are indicative steps along the path we can go and, indeed, on which we must go.

The good society cannot allow itself to be identified with the nation-state alone; it must recognize and support the larger international forces to which the individual country is subject. This is not a matter of choice; it is the modern imperative.

16

The Poor of the Planet I

THE SHAPING HISTORY

I T HAS BY NOW been adequately urged that the problems and possibilities of the human race do not respect national frontiers; in a civilized society there is concern for the world as a whole. And there must be special concern for the millions and hundreds of millions who live outside the boundaries of the more fortunate nations. These, to repeat, are people too. To frame this obligation, one must turn at least briefly to the relevant history.

There was a time when what are now the rich and developed countries ruled what are now the poor lands. Thus, Britain, France, Belgium, the Netherlands, Germany, Italy, Russia and, with a special energy and geographic scope, Spain and Portugal were all colonial powers. As was, if briefly, the United States. The memory of this rule is still vivid in the poor countries; so are the consequences of its coming to an end.

That rulers and ruled had radically different views of imperial power is taken for granted. The former thought well of their own strength and influence in the colonial world; from the latter there was resentment leading on, in the frequent case, to quiet or open revolt.

It is not absolutely clear that the adverse reaction on the part of the ruled to the power of the ruler was always quite as constant or complete as is now commonly assumed. The Roman Empire, the foremost example of imperial authority, was accepted by many of its subject peoples, a great number of whom thought it better to be within than without. Nor can one doubt that Rome, as Greece before it and as Spain, Britain, France and other imperial powers later, had a civilizing role and provided patterns of culture, government and law that were enduring contributions to the lands they had conquered. That, however, was not enough.

In the second half of the twentieth century, there came the greatest change — revolution is not too strong a word — in some thousands of years. The extent and depth of this change can only arouse wonder: imperialism, colonialism as anciently it had existed, came everywhere abruptly to an end. Suddenly in Asia, Africa, the South Pacific, colonial rule was a thing of the past. The right of some countries to govern others was no longer approved either in reality or in law. Self-government, respect for sovereign power, became accepted.

Looking back, it is still hard to imagine so comprehensive a change in so little time. The very word "imperialism" acquired a strong overtone of condemnation. As did "colonialism." This was true not alone in the former colonies; they became terms of disrepute in the former imperial or colonial powers as well. The whole world heralded a new enlightenment. A further and final act of decolonization occurred in the years after 1989, when the nations of Eastern Europe were released from Soviet influence and the former Soviet Union itself was dissolved.

All this, to repeat, was universally welcomed. The escape from imperial power, the reality of political independence, were seen and accepted as a social and political good. Imperial power is now without defenders. This, the good society approves. But first a word on why imperial power came so suddenly, so dramatically, to an end. And whether, as widely supposed, it still continues in more disguised but not necessarily less effective forms. From this comes the present state of the once-colonial world and the obligation of the good society to the poor of the planet.

The people of the former colonial regimes, it cannot be doubted, actively sought their independence. In important cases, they ceased to be governed because they had become ungovernable by outside authority. Self-assertion, self-determination, were too strong. This was undoubtedly true in India, where, under the inspired leadership of Mohandas Gandhi and Jawaharlal Nehru, dissent had a sophistication and a resulting force far in excess of what could be mobilized in opposition.

In Indochina and Algeria against the French, Angola and Mozambique against the Portuguese, there was organized military resistance to the colonial power. Remarkably, however, in much of the colonial world the erring brothers, as with regret they were regarded by many, were allowed to go in peace. In the United States there was no thought of keeping the Philippines by force. Or the other, lesser territories of the small American empire. In most of Africa, much of Asia and elsewhere there was no effective resistance to, or by, the colonial power. Colonial rule peacefully, inevitably, ceased to exist. The colonial peoples celebrated their new freedom, while the former rulers celebrated their wise acceptance of the new and more civilized reality. In Britain, the United States, France, the Netherlands, Belgium — the imperial powers, great and small — there was political support and applause. Decolonization was seen as a triumph of good over evil, a defeat for the forces of political obsolescence and reaction.

The self-approval was excessive. There was another and more influential circumstance at work here: colonialism no longer served any important economic interest. Indeed, it was possible that it now existed at some net cost. Economic advantage joined hands with idealism; this is a coalition that is always a vital force for social change.

Once, in the days of landlords and landed interest and merchants and merchant interest, colonialism had a powerful economic base. The acquisition of contiguous or even distant landed territory brought revenues and exploitable peasant manpower. It is for this reason that the more vulnerable military mind is still fixed firmly on the defense of frontiers. Territory is sacred; what else could be more important?

More important, in fact, was what benefited the merchant and industrial interests. Merchant and early industrial capitalism centered on the procurement of raw materials, tropical products, exotic handicrafts and elementary manufactures from the colonial lands and the return of industrial factory products thereto. With colonial possession went a national monopoly or near monopoly of this trade. And in the governments of the colonial powers the traders and manufacturers spoke with a strong political voice on their own behalf. This voice and that of the government were often identical.

By the end of World II, and for some time before, the merchant and elementary industrial or factory interests had diminished to a negligible, even archaic role. Economic development was now centered internally, not externally; it was from domestic economic growth that nations now prospered and were rewarded. Trade between the industrial countries was dominant; economic relations with the colonial world were now marginalized. Lenin, with undoubted exaggeration, once averred that the workers of the advanced capitalist lands lived on the backs of the colonial masses. No one could imagine that this was any longer even remotely true. It has been estimated that the loss by the Netherlands of its great Indonesian possessions after World War II was compensated for by a mere two years of domestic economic growth.

Thus the colonies could depart without economic cost.

Few, certainly not many, in the United States suffered
financial loss from the liberation of the Philippines. Had
there been a strong economic interest, there would have
been a powerful lobby expressing that interest, and the
result would have been quite different. Certainly the passage would not have been so quiet, so peaceful.

There is, however, the second question: whether a more
subtle, more sophisticated form of imperialism emerged
to rule. Did one kind of colonial rule give way to another?

That a new form of external control was replacing the old
imperialism was very much in mind in many of the old
colonial lands. Instead of government-sponsored imperialism, there was thought now to be privately sponsored imperialism, its visible instrument the transnational corporation. The former colonial land needed, in consequence
and as a very practical matter, to keep a close eye on foreign corporate activity and investment, for they were the
new imperial threat. This had an important effect on attitudes toward economic development in the new states.

Such a possibility may now be dismissed, and, indeed,
increasingly it is. The political power and influence of the
transnational corporation and of those associated generally with foreign investment were greatly overestimated.
They derived from the mystique of capitalism, not from
its reality. In practical terms, serious interference in local
politics by an international enterprise was too readily apparent, too likely to be counterproductive.

More important was the changed character of the great

business corporation itself. Originally it was the manifestation of capitalist power; it is now perceived, as an earlier chapter has told, as a large, sometimes immobile, bureaucracy. Concern for its omnipotence has given way, in the frequent case, to fear of its ineptitude. Once, indeed, there was the United Fruit Company in the banana republics of Central America. And the great oil companies dominant in the Middle East. Now no longer. Instead, there is bureaucratic caution. And the modern reality is that the former colonial countries now welcome foreign investors without fear; foreign investment is something not to be resisted but to be sought. This is notably true in India, at one time the most sensitive of the new nation-states. It has become clear that corporate power, economic power, is not the expression of a new imperialism.

In the years following World War II and until very recent times, there was, however, a more compelling neo-imperialist form. What emanated from the Soviet Union and in some measure China and from the United States and to a lesser degree Western Europe was the colonial expression of the Cold War. There was the strong Soviet influence and control in Eastern Europe. There was the hope in the Soviet Union and the paranoiac fear in the United States that the less developed lands of the planet would make Communism, not capitalism, their approved choice. The extension of superpower influence to the new and poorer nations was thus seen as the new form of imperialism. In one of the most damaging errors of modern times it was believed that there could be socialism before capitalism, that the incredibly complex administrative tasks of comprehensive central planning and control could be assumed by simple peasant people. This error had a forthright and disastrous military aspect in Afghanistan, as it did beyond reason in Indochina, particularly Vietnam.

But again this is now history. The breakup of the Soviet Union, the downfall of Communism and the end of the Cold War brought this rule of error to an end. Accordingly, in the decade of the 1990s, for the first time in history recent and ancient, there is no tangible manifestation of imperialism readily to be seen. There are great economic powers and lesser ones. There is varying military strength, much of it, as already demonstrated, of uncertain purpose. Imperialism, colonialism, belong to the past. We speak, sometimes reflectively, of the end of history; here, indeed, history has come to an end.

The good society must accept that in the relations between rich countries and poor, between the former colonial powers and their colonies, the world and the human situation have changed for all time. The foreign policy of the good society must still be sensitive to the past and particularly to anything that may seem to suggest a resurgent exercise of colonialism.

This is especially necessary in the case of the United States. The largest and militarily the most dominant of the former imperial powers or of those so regarded, it naturally arouses the greatest fear of some imperial residuum. This is heightened by the frequent and often ill-considered reference to the need of the United States to exercise leadership in the world community, to assume its *natural* leadership role. Caution and restraint are here of prime importance. Leadership, initiative, are, of course, still required, but in the modern world they must be a wise and normally a collective response to need, not a seeming manifestation of imperial right.

The fortunate countries must now deal with the imperial legacy — the grave, indeed intolerable, human suffering left in its wake. This is equally important for the good society, as will be discussed in the next chapter.

17

The Poor of the Planet II

WHAT THE GOOD SOCIETY

MUST DO

THE PROBLEMS afflicting the poor of the planet do not end with the history just recounted. When the former colonial possessions achieved independence, they were forced to take on the most demanding of human tasks, the provision of honest, reliable and responsible government. Many have failed. From this, in turn, has come economic failure, for economic success depends on the support and supervision of a stable, efficient, effective governmental structure. Without it, the most essential of the requirements for economic development is unrealized. In the last century, as already indicated, a question as to what economic progress demanded would have brought a prompt reply in the United States, as also in Europe: good government, good education and possibly good transportation. This is still a controlling rule in our own time.

That the nation-state in the good society has a basic responsibility for its own people is not in doubt. But no country can be comfortable and content in its affluence if others are abjectly poor. And certainly not if, as in the case of the former colonial powers, it had once had responsibility for the less fortunate people; its obligation did not end when its colonies assumed self-rule. In most of Africa, much of Asia, a good part of Latin America, stark poverty is still endemic. The good society cannot set itself apart from this poverty; it must be on the conscience of all, its elimination part of the public policy of all. The nation-state must not attempt to evade responsibility by accepting the most commonly used formula for selfishness and self-interest: "That's another country; they are not our problem."

In the last fifty years the poor lands, the former colonies, have, indeed, received more than slight attention from the more fortunate. Some of this was the result of the hope or fear of Communism. More benign and intelligent was the operative role of compassion, the sense of obligatory concern. This was not negligible. An influential constituency in the rich countries has consistently expressed sympathy and supported aid for those who are poor, as have the World Bank and lesser international agencies. The good society strongly supports this broad effort, but it also requires that the aid be effectively dispensed and employed. This has not always, perhaps not usually, been the case.

In the early years of assistance after World War II and the decolonization that then occurred, the new countries and the old united in the belief that escape from poverty into successful economic development involved primarily the transfer of the heavy furniture of the developed economy. To the developing lands, as optimistically they were called, went the steel mills, electrical generating plants, chemical plants, machine tool plants that were so prominent a feature of the mature economy. This was thought to be economic progress, an end to poverty. It was, in fact, a major error, an idle dream. Ignored or passed over were the two greater needs previously mentioned: a stable political order and general education for the masses. The steel mills, hydroelectric plants and shiny airports, now sited among ignorant people, became sterile monuments to error — and failure. In more recent times this has been in some measure recognized, as the role of stable, effective, honest government has become evident and accepted and the importance of education has come to be realized. On this a later word.

Also in the early days of development assistance, agriculture was frequently neglected; emphasis was on the cities and their inhabitants. That was where development occurred. Food prices were often fixed to favor the urban proletariat, and this had a depressive effect on agricultural production. To this last policy, however, there have been some obvious exceptions, the leading example, as so often, being in India. Since Independence, the Indian population has more than doubled, but, supported by grain hybrids, fertilizer, irrigation, other soil and water management and assured prices for farmers, so has food production.

*

However, in much of the former colonial world government still functions in an economically and socially adverse way. Instability, incompetence, corruption and the dictatorship of the favored few are all too frequent. In extreme but not exceptional instances there is civil disorder and conflict. Even where things are better, the routine tasks of the state — the collection of taxes, the rendering of essential services, the provision of a firm legal basis for economic progress — are badly performed or not performed at all. A damaging commerce develops in the sale by politicians and public servants of exceptions or privileged interpretations of the laws and regulations. This is true even when the country is relatively well governed, as in India. In the deeper recesses of Africa it is not similarly a problem, because there no one imagines that the regulations will be enforced.

From the foregoing comes the role of the good society. It begins with generosity — financial help based not on political or economic self-interest but, as has been sufficiently stressed, on concern for fellow human beings. Where there is internal peace and passably effective government, the primary emphasis must be on education. For that, money must be generously available — money for schools, equipment and teachers, and especially for the training of teachers. Capital, the basic requirement for material investment, moves readily from one country to another. Teachers, the essential instruments of educational progress, are less mobile. An international teacher training corps — teachers of teachers — is one of the major needs of economic development. More generally, education must be central to policy in all fields. I here urge again a point elsewhere made: in this world there is no literate population that is poor, no illiterate population that is not. Given an educated population, economic advance becomes, in some measure, inevitable. Only then comes the truly effective use of more general development aid.

In the poor countries the breakdown of law and order is also a frequent and especially urgent problem. This has been particularly the case in Africa, but it has occurred from time to time in Central America, Asia and in Europe. Liberia, Somalia, Rwanda, Nicaragua, Haiti, Bosnia, are recent examples as this is written. Hundreds of thousands of their citizens have died either in the resulting conflict or from the ensuing deprivation and forced migration.

It has long been recognized and accepted that it is not within the sovereign right of any nation-state to attack another. International law forbids it; the United Nations exists largely to prevent it. The slaughter by any country of its own people is a different case. It invites disapproval, condemnation, but it does not justify the kind of response from other countries that, for example, followed the takeover of Kuwait by Iraq. Nonetheless, the human suffering and economic and social devastation from internal conflict can be, and in recent times have been, far greater than those from international conflict. This must be understood; there must be action by the good society against these most appalling of human tragedies. The need must be recognized, accepted, and the response given a substantial, predictable form.

In these last years in Somalia, Rwanda, earlier by its neighbors in Liberia and in the former Yugoslavia there has been international intervention to still domestic violence or to feed, shelter, extend medical care to and otherwise succor those afflicted. Each case has been regarded as exceptional; action has been thought an accommodation to seemingly special circumstance. This must not be so in the future; the breakdown of law and order and the associated human suffering must be seen as a wholly predictable event in the poor lands. And so must intervention — the setting aside of sovereignty to rescue and protect distressed and endangered populations.

The reference has been to international action; this means, effectively, the United Nations. In both international opinion and international law unilateral action by any one state is strongly suspect. This is especially so if it is by a major power such as the United States; here the enduring specter of imperialism. Accordingly, action must always be internationally sanctioned and under international control. In the future this must be seen as one of the most important functions of the United Nations, supported according to ability to pay by the rich lands and, quite definitely, by the United States. Dispatch of the requisite police *cum* military personnel must be a general and accepted obligation. The present American position

— that even though the United States has the world's strongest and best-endowed military establishment, it must not risk the political consequences at home of incurring any casualties abroad — must especially be revised. The bleak course of history has already defined the task; the good society must respond.

There has been emphasis in this chapter on the role of compassion — the obligation of the fortunate to the deprived. With the decline and disappearance of the anti-Communist fear and paranoia that once justified aid from the rich to the poor, this must now be a principal sustaining factor for such aid. But it is not the only one. There is advantage, even safety, for all peoples if the world is tranquil, at peace. As has been sufficiently emphasized, the poverty of nations is an instigating and nurturing source of conflict, and in the modern world the fortunate lands are at peace with each other and at peace with themselves. International comity will be served and will only be served by the creation of economic and social well-being in all countries. Conflict, as all know, is an infection that can spread. The assistance and the acceptance of the larger responsibility for internal tranquillity and order that are here urged will contribute to peace for the poor lands and peace of mind for the more fortunate. Compassion has a human face, but it also renders a very practical service.

Not all of the nations needing help in the world are the traditionally poor. In the years since 1989, the countries of

Eastern Europe and of the former Soviet Union have been making the difficult passage from comprehensive socialism to a market economy. And to democracy and peaceful membership in the larger family of nations. Here too there is need for intelligent, self-serving generosity, for this transition is difficult and has an adverse bearing on many people. Such adversity, in turn, is an enormous threat to democracy, but outside assistance can ease it in multiple ways. From this will then come the promise of stable international relations and the lifting of the fear of international conflict.

The concern that has been urged for the poor of the planet and the acceptance of the cost that would be entailed in going to their aid must extend equally to those endangered by the great transition in Eastern Europe. The good society cannot live under the shadow of social disorder in the former Communist lands and the resulting threat of warfare and even nuclear devastation. As this is written, time here is very short, action too long delayed.

18

The Political Context

BOOKS OF THIS MOOD and genre, as I have previously had occasion to suggest, almost always end on the same note. Having defined what is good and achievable, they assume that the necessary political response will follow, if not soon, then in time. People have an instinct, immediate or eventual, for what is right. Having specified this, the writer's task is done; action proceeds from there. It is this optimism that sustains the toil of thought and authorship; thus is demonstrated the ultimate power of the ideas that John Maynard Keynes, in his most famous statement, said ruled a world that is ruled by little else.

There is no such fragile optimism here. Because in the modern polity there are two groups that are unequal in power and influence, democracy has become an imperfect thing. On the one hand, as we have seen, there are the favored, the affluent and the rich, not excluding the corporate bureaucracy and the business interest, and on the other, the socially and economically deprived, along with the considerable number who, out of concern and compassion, come to their support. It can be, and most clearly is, an unequal contest.

The decisive step toward a good society is to make democracy genuine, inclusive. As this is written, a sharp debate is in progress in the United States over welfare reform; proposed, in essence, is a partial withdrawal of the safety net that protects the poorest citizens and especially their children. Without it they would be condemned to hunger, otherwise treatable illness and the discomforts of inadequate housing; they would feel they were being further denigrated as inherently inferior.

Things would be much changed if the less fortunate and the poor resorted reliably to the ballot box to redress their ills. Their votes would then be specifically, even diligently, solicited, with emphasis on the need for the safety net and on the quality of publicly provided schools, housing, health care, recreational facilities and much else. In the political turnover in the United States in the autumn of 1994, as previously indicated, those opposing aid to the poor in its several forms won their stunning victory with the support of less than one quarter of all eligible voters, fewer than half of whom had gone to the polls. The popular and media response was that those who had prevailed represented the view and voice of the public. Had there been a full turnout at the election, both the result and the reaction would have been decidedly different. The sense of social responsibility for the poor would have been greatly enhanced.

Thus the prime essential for achieving the good society

is a more nearly perfect expression of democratic will; democracy must be made genuine, inclusive. In the state of New Jersey in a recent gubernatorial election the level of taxation agreeable to the reasonably affluent and the rich and the promise of tax reduction were the major (and successful) issues. A highly placed and more than adequately self-confident political operative told afterward with satisfaction of how he had distributed money to the pastors of black churches as payment for not encouraging their poor parishioners to vote. He later denied having done what he praised himself for doing, a difficult procedure, but the exercise against democratic process was a trifle too obvious. Nonetheless, his political instincts were wholly accurate. The votes of the poor are essential for getting the public services they need, for raising the necessary revenues and for instituting the broad policies that ameliorate poverty — in short, for achieving a first step toward the good society.

Democracy has its compelling requirements. There must be a clear perception of the goals to which the majority is, or should be, committed, goals that this essay has sought to define. And there must be organization to mobilize voters and persuade legislators and Presidents in support of those goals. In recent times the nature and magnitude of this effort have become wonderfully clear.

Money, voice and political activism are now extensively controlled by the affluent, the very affluent and the business interests, and to them much political talent is inevitably drawn. The expression of their goals is then accepted as public opinion and, a significant point, is so designated by the media every day. In the United States the Republican Party is avowedly on the side of the fortunate, and to the influence and wealth of the latter the Democratic Party, or many of its members, are also attracted. The result, or at a minimum the possibility, is a two-party system in which both parties respond in policy and action to the needs and desires of the well- and richly endowed.

In the good society voice and influence cannot be confined to one part of the population. In the United States the only solution is more active political participation by a coalition of the concerned and the poor. And their instrument must be the Democratic Party, for this has been its past role and the source of its past success. It has traditionally spoken for effective action by the state on behalf of the less advantaged when that was required. And it has resisted the currently avowed tendency to identify government as a burden when it comes to the aid of the poor but not when the needs or preferences of the affluent are being addressed.

"Practical" politics, it is held, calls for policies that appeal to the fortunate. The poor do not vote; the alert politician bids for the comfortable and the rich. This would be politically foolish for the Democratic Party; those whose primary concern is to protect their income, their capital and their business interest will always vote for the party that most strongly affirms its service to their pecuniary well-being. This is and has always been the Republicans. The Democrats have no future as a lower-grade substitute.

Nor is there a substantial possibility that a new party will be established and succeed in the United States, although that has been the hope of some of the underprivileged in the past. It would encounter an institutional structure, the two-party system, that over a century and much more has been a stable feature of American government. Two parties are entrenched in thought and action, and so they will continue to be.

In other industrial countries the situation is, on the whole, more favorable for the poor. There, a larger part of the electorate goes to the polls; in an astonishingly forthright response to political domination by the affluent, some countries — Australia, Belgium — have even made voting compulsory. This is a step well beyond the achievable in the United States; perhaps it is within the sovereign, inalienable rights of the American citizen that he or she may boycott the election process. Nonetheless, the central flaw of the good society is not democracy but that democracy is imperfect. Only when all vote — all but the eccentric few — will the good society achieve its urgent goals.

It is inevitable that critics who have survived to these final pages will say as with one voice that what is here written is out of step with the times. The fortunate, including those who speak for them and those allied in politics, are securely in command. They are the political reality; so they will be for the foreseeable future.

Not necessarily. Let there be a coalition of the concerned and the compassionate and those now outside the political system, and for the good society there would be a bright and wholly practical prospect. The affluent would still be affluent, the comfortable still comfortable, but the poor would be part of the political system. Their needs would be heard, as would the other goals of the good society. Aspirants for public office would listen. The votes would be there and would be pursued. As now with the safety net, health care, the environment and especially the military power, the good society fails when democracy fails. With true democracy, the good society would succeed, would even have an aspect of inevitability.

INDEX

Index